Applied Biofluid Mechanics

Applied Biofluid Mechanics

Lee Waite, Ph.D., P.E.

Jerry Fine, Ph.D.

New York   Chicago   San Francisco   Lisbon   London   Madrid
Mexico City   Milan   New Delhi   San Juan   Seoul
Singapore   Sydney   Toronto

The McGraw·Hill Companies

Library of Congress Cataloging-in-Publication Data

Waite, Lee.
   Applied biofluid mechanics / Lee Waite, Jerry Fine.
      p.   cm.
   Includes bibliographical references.
   ISBN-13: 978-0-07-147217-3 (alk. paper)
   ISBN-10: 0-07-147217-7 (alk. paper)
   1. Blood flow.   2. Fluid mechanics.   3. Blood—Circulation.
   4. Biomedical engineering.   I. Fine, Jerry Michael.   II. Title.
   [DNLM: 1. Cardiovascular Physiology.   2. Biomedical Engineering.
   3. Body Fluids—physiology.   4. Respiratory Physiology.   WG 102 W145a
   2007]
   QP105.W346   2007
   610.28—dc22   2007013612

1 2 3 4 5 6 7 8 9 0   DOC/DOC   0 1 3 2 1 0 9 8 7

ISBN-13: 978-0-07-147217-3
ISBN-10: 0-07-147217-7

Sponsoring Editor: Stephen S. Chapman
Production Supervisor: Pamela A. Pelton
Editing Supervisor: Stephen M. Smith
Project Manager: Rasika Mathur
Copy Editor: Donna Coggshall
Proofreader: Yumnam Ojen
Indexer: Robert Swanson
Art Director, Cover: Jeff Weeks
Composition: International Typesetting and Composition

Printed and bound by RR Donnelley.

McGraw-Hill books are available at special quantity discounts to use as premiums and sales promotions, or for use in corporate training programs. For more information, please write to the Director of Special Sales, McGraw-Hill Professional, Two Penn Plaza, New York, NY 10121-2298. Or contact your local bookstore.

This book is printed on acid-free paper.

ABOUT THE AUTHORS

LEE WAITE, PH.D., P.E., is Head of the Department of
Applied Biology and Biomedical Engineering, and Director
of the Guidant/Eli Lilly and Co. Applied Life Sciences
Research Center, at Rose-Hulman Institute of Technology in
Terre Haute, Indiana. He is also the author of *Biofluid
Mechanics in Cardiovascular Systems*, published by
McGraw-Hill.

JERRY FINE, PH.D., is Associate Professor of Mechanical
Engineering at Rose-Hulman Institute of Technology.
Before he joined the faculty at Rose, Dr. Fine served as
a patrol plane pilot in the U.S. Navy and taught at the
U.S. Naval Academy.

Contents

Preface

Biomedical engineering is a discipline that is multidisciplinary by definition. The days when medicine was left to the physicians, and engineering was left to the engineers, seem to have passed us by. I have searched for an undergraduate/graduate level biomedical fluid mechanics textbook since I began teaching at Rose-Hulman in 1987. I looked for, but never found, a book that combined the physiology of the cardiovascular and pulmonary systems with engineering of fluid mechanics and hematology to my satisfaction, so I agreed to write a monograph. Ken McCombs at McGraw-Hill was satisfied well enough with that work that he asked me to write the textbook version that you see here. *Applied Biofluid Mechanics* includes problem sets and a solutions manual that traditionally accompany engineering textbooks.

   Applied Biofluid Mechanics begins in Chapter 1 with a review of some of the basics of fluid mechanics, which all mechanical or chemical engineers would learn. It continues with two chapters on cardiovascular and pulmonary physiology followed by a chapter describing hematology and blood rheology. These five chapters provide the foundation for the remainder of the book, which focuses on more advanced engineering concepts. Dr. Fine has added some particularly nice improvements in Chapters 7 and 10 concerning solutions for, and modeling of, pulsatile flow.

   My 10-week, graduate-level, biofluid mechanics course forms the basis for the book. The course consists of 40 lectures and covers most, but not all, of the material contained in the book. The course is intended to prepare students for work in the health care device industry and others for graduate work in biomedical engineering.

   In spite of great effort on the part of many proofreaders, I expect that mistakes will appear in this book. I welcome suggestions for improvement from all readers, with intent to improve subsequent printings and editions.

LEE WAITE, PH.D., P.E.

Acknowledgments

I would like to thank my BE525 students from fall term of 2006, who through their proofreading helped to improve this book, *Applied Biofluid Mechanics*. I wish to thank Steve Chapman and the editorial and production staff at McGraw-Hill for their assistance.

Special thanks go to my friend and co-author Dr. Jerry Fine, without whom it would have been extremely difficult to meet the agreed-upon deadlines for this book. Dr. Fine is especially responsible for significant improvements in Chapters 7 and 10.

Jerry would like to give a special acknowledgment to his father, Dr. Neil C. Fine, for his encouragement over the years, and for being a superb example of a life-long learner.

Thanks once again to the faculty in the Applied Biology and Biomedical Engineering Department at Rose-Hulman for their counsel and for putting up with a department head who wrote a book instead of giving full concentration to departmental issues.

In writing this book I have been keenly aware of the debt I owe to Don Young, my dissertation advisor at Iowa State, who taught me much of what I know about biomedical fluid mechanics.

Most of all I would like to thank my colleague, wife, and best friend Gabi Nindl Waite, who put up with me during the long evening hours and long weekends that it took to write this book. Thanks especially for everything you taught me about physiology.

LEE WAITE, PH.D., P.E.

Applied Biofluid Mechanics

Review of Basic Fluid Mechanics Concepts

1.1 A Brief History of Biomedical Fluid Mechanics

People have written about the circulation of blood for thousands of years. I include here a short history of biomedical fluid mechanics, because I believe it is important to recognize that, in all of science and engineering, we "stand on the shoulders of giants."[1] In addition, it is interesting information in its own right. Let us begin the story in ancient times.

The Yellow Emperor, Huang Ti, lived in China from about 2700 to 2600 BC and, according to legend, wrote one of the first works dealing with circulation. Huang Ti is credited with writing *Internal Classics,* in which fundamental theories of Chinese medicine were addressed, although most Chinese scholars believe it was written by anonymous authors in the Warring Period (475–221 BC). Among other topics, *Internal Classics* includes the Yin-Yang doctrine and the theory of circulation.

Hippocrates (Fig. 1.1), who lived in Greece around 400 BC, is considered by many to be the father of science-based medicine and was the first to separate medicine from magic. Hippocrates declared that the human body was integral to nature and was something that should be understood. He founded a medical school on the island of Cos, Greece, and developed the Oath of Medical Ethics.

---

[1] "If I have seen further (than others), it is by standing on the shoulders of giants." This was written by Isaac Newton in a letter to Robert Hooke, 1675.

Figure 1.1 Hippocrates. *Courtesy of the National Library of Medicine Images from the* History of Medicine, *B029254.*

Aristotle, a highly influential early scientist and philosopher, lived in Greece between 384 and 322 BC. He wrote that the heart was the focus of blood vessels, but did not make a distinction between arteries and veins.

Praxagoras of Cos was a Greek physician and a contemporary of Aristotle. Praxagoras was apparently the first Greek physician to recognize the difference between arteries (carriers of air, as he thought) and veins (carriers of blood), and to comment on the pulse.

The reasoning behind arteries as carriers of air makes sense when you realize that, in a cadaver, the blood tends to pool in the more flexible veins, leaving the stiffer arteries empty.

In this book, we will also consider the mechanics of breathing at high altitudes, in the discussion of biomedical fluid mechanics. Let us turn our history in that direction. It is thought that Aristotle (384–322 BC) was aware that the air is "too thin for respiration" on top of high mountains.

Francis Bacon (1561–1626) in his *Novum Organun,* which appeared in 1620, includes the following statement (Bacon, 1620, pp. 358–360) from an English translation *of* the Latin text. "The ancients also observed, that *the* rarity of the air on the summit of Olympus was such that those who ascended it were obliged to carry sponges moistened with vinegar and water*, and* to apply them now and then to their nostrils, as the air was not dense enough for their respiration."

There is much evidence that the ancients seem to have known something about the mechanics of respiration at high altitudes. The following colorful description of mountain sickness comes from a classical Chinese history of the period preceding the Han dynasty, the Ch'ien Han Shu. The Chinese text is from Pan Ku (AD 32–92), and the English translation is from Alexander Wylie (1881) as appears in John West's *High Life* (1998). Speaking of a journey through the mountains, the text comments:

> Again, on passing the Great Headache Mountain, the Little Headache Mountain, the Red Land, and the Fever Slope, men's bodies become feverish, they lose color, and are attached with headache and vomiting; the asses and cattle being all in like condition. Moreover there are three pools with rocky banks along which the pathway is only 16 or 17 inches wide for a length of some 30 *le*, over an abyss...

The first description of high-altitude pulmonary edema appeared some 400 years later. Fâ-hien was a Chinese Buddhist monk who undertook an amazing trip through western China, Sinkiang, Kashmir, Afghanistan, Pakistan, and northern India to Calcutta by foot. He continued the trip by boat to Sri Lanka and then on to Indonesia and finally to the China Sea ending in Nanjing. The journey took 15 years from AD 399 to 414. While crossing the "Little Snowy Mountains" in Afghanistan, his companion became ill.

Having stayed there till the third month of winter, Fâ-hien and two others, proceeding southward, crossed the Little Snowy Mountains, on which the snow lies accumulated both in winter and in summer. On the northern side of the mountains, in the shade, they suddenly encountered a cold wind which made them shiver and become unable to speak. Hwuyking could not go any further. White froth came down from his mouth, and he said to Fâ-hien, "I cannot live any longer. Do you immediately go away, that we do not all die here." With these words, he died. Fâ-hien stroked the corpse and cried out piteously, "Our original plan has failed; it is fate. What can we do?"

Father Joseph de Acosta (1540–1600) was a Spanish Jesuit priest who left Spain and traveled to Peru in about 1570. He wrote the book *Historia Natural y Moral de las Indias,* which was first published in Seville in Spanish in 1590. Acosta had become a Jesuit priest at the age of 13. As he left Spain, he traveled across the Atlantic to Nombre de Dios, a town on the Atlantic coast of Panama, near the mouth of the Río Chagres, near

Figure 1.2 Authors: Lee Waite and Jerry Fine, on the equator in Ecuador (Mt. Cayambe, 5790 m, near the highest point on the equator), agreeing with Father Joseph de Acosta who laughed, 400 years earlier, over Aristotle's prediction of unbearable heat in the "burning zone."

Colon, and then journeyed through 18 leagues of tropical forest to Panama. West writes

> From Panama he embarked for Peru with some apprehension because the ancient philosophers had taught that the equator was in the "burning zone" where the heat was unbearable. However he crossed the equator in March, and to his surprise, it was so cold that he was forced to go into the sun to get warm, where he laughed at Aristotle and his philosophy.

See Fig. 1.2. (Acosta, 1604, p. 90.)

On visiting a Peruvian mountain, which he called, "Pariacaca," which is thought by some to be the mountain called today Tullujuto, Acosta wrote

> For my part I holde this place to be one of the highest parts of land in the worlde; for we mount a wonderfull space. And in my opinion, the mountaine *Nevade of Spaine*, the *Pirenees*, and the *Alpes of Italie*, are as ordinarie houses, in regard of hie Towers. I therefore perswade my selfe, that the element of the aire is there so subtile and delicate as it is not proportionable with the breathing of man, which requires a more grosse and temperate aire, and I beleeve it is the cause that doth so much alter the stomacke, & trouble all the disposition.

This account of mountain sickness and the thinness of the air at high altitudes is perhaps the most famous from this time period (Acosta, 1604, Chapter 9, pp. 147–148).

Figure 1.3 William Harvey. *Courtesy of the National Library of Medicine Images from the* History of Medicine, *B014191.*

The culmination of our history is the story of how the circulation of blood was discovered. William Harvey (Fig. 1.3) was born in Folkstone, England, in 1578. He earned a BA degree from Cambridge in 1597 and went on to study medicine in Padua, Italy, where he received his doctorate in 1602. Harvey returned to England to open a medical practice and married Elizabeth Brown, daughter of the court physician to Queen Elizabeth I and King James I. Harvey eventually became the court physician to King James I and King Charles I.

In 1628, Harvey published, "An anatomical study of the motion of the heart and of the blood of animals." This was the first publication in the Western World that claimed that blood is pumped from the heart and recirculated. Up to that point, the common theory of the day was that food was converted to blood in the liver and then consumed as fuel. To prove that blood was recirculated and not consumed, Harvey showed, by calculation, that blood pumped from the heart in only a few minutes exceeded the total volume of blood contained in the body.

Jean Louis Marie Poiseuille was a French physician and physiologist, born in 1797. Poiseuille studied physics and mathematics in Paris. Later, he became interested in the flow of human blood in narrow tubes. In 1838, he experimentally derived and later published Poiseuille's law. Poiseuille's law describes the relationship between flow and pressure gradient in long tubes with constant cross section. Poiseuille died in Paris in 1869.

Otto Frank was born in Germany in 1865, and he died in 1944. He was educated in Munich, Kiel, Heidelberg, Glasgow, and Strasburg. In 1890, Frank published, "Fundamental form of the arterial pulse," which contained his "Windkessel theory" of circulation. He became a physician in 1892 in Leipzig and became a professor in Munich in 1895. Frank perfected optical manometers and capsules for the precise measurement of intracardiac pressures and volumes.

1.2   Fluid Characteristics and Viscosity

A fluid is defined as a substance that deforms continuously under application of a shearing stress, regardless of how small the stress is. Blood is a primary example of a biological fluid. To study the behavior of materials that act as fluids, it is useful to define a number of important fluid properties, which include density, specific weight, specific gravity, and viscosity.

Density is defined as the mass per unit volume of a substance and is denoted by the Greek character ρ (rho). The SI units for ρ are kg/m^3, and the approximate density of blood is 1060 kg/m^3. Blood is slightly denser than water, and red blood cells in plasma[2] will settle to the bottom of a test tube, over time, due to gravity.

Specific weight is defined as the weight per unit volume of a substance. The SI units for specific weight are N/m^3. Specific gravity s is the ratio of the weight of a liquid at a standard reference temperature to the weight of water. For example, the specific weight of mercury $S_{Hg} = 13.6$ at 20°C. Specific gravity is a unitless parameter.

Density and specific weight are measures of the "heaviness" of a fluid, but two fluids with identical densities and specific weights can flow quite differently when subjected to the same forces. You might ask, "What is the additional property that determines the difference in behavior?" That property is viscosity.

---

[2] Plasma has a density very close to that of water.

1.2.1   Displacement and velocity

To understand viscosity, let us begin by imagining a hypothetical fluid between two parallel plates which are infinite in width and length. See Fig. 1.4.

The bottom plate A is a fixed plate. The upper plate B is a moveable plate, suspended on the fluid, above plate A, between the two plates. The vertical distance between the two plates is represented by h. A constant force F is applied to the moveable plate B causing it to move along at a constant velocity V_B with respect to the fixed plate.

If we replace the fluid between the two plates with a solid, the behavior of the plates would be different. The applied force F would create a displacement d, a shear stress τ in the material, and a shear strain γ. After a small, finite displacement, motion of the upper plate would cease.

If we then replace the solid between the two plates with a fluid, and reapply the force F, the upper plate will move continuously, with a velocity of V_B. This behavior is consistent with the definition of a fluid: a material that deforms continuously under the application of a shearing stress, regardless of how small the stress is.

After some infinitesimal time dt, a line of fluid that was vertical at time $t = 0$ will move to a new position, as shown by the dashed line in Fig. 1.4. The angle between the line of fluid at $t = 0$ and $t = t + dt$ is defined as the shearing strain. Shearing strain is represented by the Greek character γ (gamma).

The first derivative of the shearing strain with respect to time is known as the rate of shearing strain $d\gamma/dt$. For small displacements, $\tan(d\gamma)$ is approximately equal to $d\gamma$. The tangent of the angle of shearing strain can also be represented as follows:

$$\tan(d\gamma) = \frac{\text{opposite}}{\text{adjacent}} = \frac{V_B dt}{h}$$

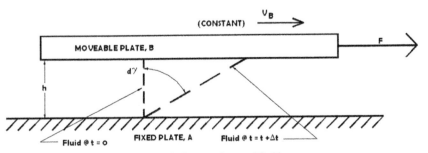

Figure 1.4 Moveable plate suspended over a layer of fluid.

Therefore, the rate of shearing strain $d\gamma/dt$ can be written as

$$d\gamma/dt \;=\; V_B/h$$

The rate of shearing strain is also denoted by $\dot{\gamma}$, and has the units of 1/s.

The fluid that touches plate A has zero velocity. The fluid that touches plate B moves with the same velocity as that of plate B, V_B. That is, the molecules of the fluid adhere to the plate and do not slide along its surface. This is known as the no-slip condition. The no-slip condition is important in fluid mechanics. All fluids, including both gasses and liquids, satisfy this condition.

Let the distance from the fixed plate to some arbitrary point above the plate be y. The velocity V of the fluid between the plates is a function of the distance above the fixed plate A. To emphasize this we write

$$V = V(y)$$

The velocity of the fluid at any point between the plates varies linearly between $V = 0$ and $V = V_B$. See Fig. 1.5.

Let us define the velocity gradient as the change in fluid velocity with respect to y.

$$\text{Velocity gradient} \equiv dV/dy$$

The velocity profile is a graphical representation of the velocity gradient. See Fig. 1.5. For a linearly varying velocity profile like that shown in Fig. 1.5, the velocity gradient can also be written as

$$\text{Velocity gradient} = V_B/h$$

1.2.2   Shear stress and viscosity

In cardiovascular fluid mechanics, shear stress is a particularly important concept. Blood is a living fluid, and if the forces applied to the fluid

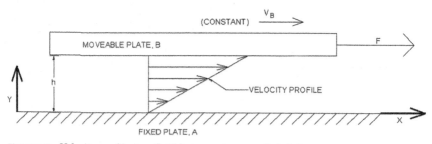

Figure 1.5   Velocity profile in a fluid between two parallel plates.

are sufficient, the resulting shearing stress can cause red blood cells to be destroyed. On the other hand, studies indicate a role for shear stress in modulating atherosclerotic plaques. The relationship between shear stress and arterial disease has been studied much, but is not yet very well understood.

Figure 1.6 represents the shear stress on an element of the fluid at some arbitrary point between the plates in Figs. 1.4 and 1.5. The shear stress on the top of the element results in a force that pulls the element "downstream." The shear stress at the bottom of the element resists that movement.

Since the fluid element shown will be moving at a constant velocity, and will not be rotating, the shear stress on the element τ' must be the same as the shear stress τ. Therefore,

$$d\tau/dy = 0 \text{ and } \tau_A = \tau_B = \tau_{\text{wall}}$$

Physically, the shearing stress at the wall may also be represented by

$$\tau_A = \tau_B = \text{force/plate area}$$

The shear stress on a fluid is related to the rate of shearing strain. If a very large force is applied to the moving plate B, a relatively higher velocity, a higher rate of shearing strain, and a higher stress will result. In fact, the relationship between shearing stress and rate of shearing strain is determined by the fluid property known as viscosity.

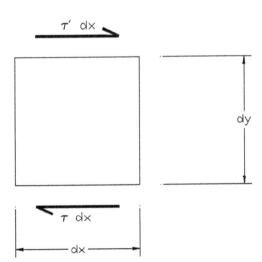

Figure 1.6 Shear stress on an element of the fluid.

1.2.3   Example problem: shear stress

Wall shear stress may be important in the development of various vascular disorders. For example, the shear stress of circulating blood on endothelial cells has been hypothesized to play a role in elevating vascular transport in ocular diseases such as diabetic retinopathy.

In this example problem, we are asked to estimate the wall shear stress in an arteriole in the retinal circulation. Gilmore et al. have published a related paper in the *American Journal of Physiology: Heart and Circulatory Physiology*, volume 288, in February 2005. In that article, the authors published the measured values of retinal arteriolar diameter and blood velocity in arterioles. For this problem, we will use their published values: 80 μm for a vessel diameter and 30 mm/s for mean retinal blood flow velocity. Later in Sec. 1.4.4, we will see that, for a parabolic flow profile, a good estimate of the shearing rate is

$$\dot{\gamma} = \frac{8V_m}{D}$$

where V_m is the mean velocity across the vessel cross section and D is the vessel inside diameter.

We will also see in the next section that the shear stress is equal to the viscosity multiplied by the rate of shearing strain, that is,

$$\tau = \mu\dot{\gamma}$$

Therefore, to estimate the shear stress on the wall of a retinal arteriole, with the data from Gilmore's paper, we can calculate

$$\tau = \frac{\mu\, 8V_m}{D} = \frac{0.0035\,\frac{Ns}{m^2}\,8(3)\,\frac{cm}{s}}{0.008\,cm} = 10.5\,\frac{N}{m^2}$$

Although 10.5 Pa seems like a low shear stress when compared to the strength of aluminum or steel, it is a relatively high shear stress when compared to a similar estimate in the aorta, 0.5 Pa. See Table 1.1.

TABLE 1.1   Estimate of Wall Shear Stress in Various Vessels in the Human Circulatory System

Vessel	ID, cm	V_m, cm/s	Shear rate[3]	Shear stress,[4] N/m^2
Aorta	2.5	48	154	0.5
Large arteriole[5]	0.05	1.4	224	0.8
Arteriole (retinal microcirculation[6])	0.008	3	3000	10.5
Capillary	0.0008	0.7	7000	24.5

Note the increasing values for shear rate and shear stress as vessel inside diameter decreases.

[3]Estimate for a parabolic flow profile, 8 times mean velocity divided by diameter.
[4]Based on a viscosity of 3.5 cP.
[5]From Milnor, 1990, p. 334.
[6]From Gilmore et al., 2005, 288: H2912–H2917.

1.2.4 Viscosity

A common way to visualize material properties in fluids is by making a plot of shearing stress as a function of the rate of shearing strain. For the plot shown in Fig. 1.7, shearing stress is represented by the Greek character τ, and the rate of shearing strain is represented by $\dot{\gamma}$.

The material property that is represented by the slope of the stress–shearing rate curve is known as viscosity and is represented by the Greek letter μ (mu). Viscosity is also sometimes referred to by the name absolute viscosity or dynamic viscosity. For common fluids like oil, water, and air, viscosity does not vary with shearing rate. Fluids with constant viscosity are known as Newtonian fluids. For Newtonian fluids, shear stress and rate of shearing strain may be related by the following equation:

$$\tau = \mu\dot{\gamma}$$

where τ = shear stress
μ = viscosity
$\dot{\gamma}$ = the rate of shearing strain

For non-Newtonian fluids, τ and $\dot{\gamma}$ are not linearly related. For those fluids, viscosity can change as a function of the shear rate (rate of shearing strain). Blood is an important example of a non-Newtonian fluid. Later in this book, we will investigate the condition under which blood behaves as, and may be considered, a Newtonian fluid.

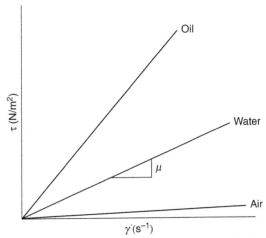

Figure 1.7 Stress versus rate of shearing strain for various fluids.

Shear stress and shear rate are not linearly related for non-Newtonian fluids. Therefore, the slope of the shear stress/shear rate curve is not constant. However, we can still talk about viscosity if we define the apparent viscosity as the instantaneous slope of the shear stress/shear rate curve. See Fig. 1.8.

Shear thinning fluids are non-Newtonian fluids whose apparent viscosity decreases as shear rate increases. Latex paint is a good example of a shear thinning fluid. It is a positive characteristic of the paint that the viscosity is low when one is painting, but that the viscosity becomes higher and the paint sticks to the surface better when no shearing force is present. At low shear rates, blood is also a shear thinning fluid. However, when the shear rate increases above 100 s^{-1}, blood behaves as a Newtonian fluid.

Shear thickening fluids are non-Newtonian fluids whose apparent viscosity increases when the shear rate increases. Quicksand is a good example of a shear thickening fluid. If one tries to move slowly in quicksand, then the viscosity is low and the movement is relatively easy. If one tries to move quickly, then the viscosity increases and the movement is difficult. A mixture of cornstarch and water also forms a shear thickening non-Newtonian fluid.

A Bingham plastic is neither a fluid nor a solid. A Bingham plastic can withstand a finite shear load and flow like a fluid when that shear stress is exceeded. Toothpaste and mayonnaise are examples of Bingham plastics. Blood is also a Bingham plastic and behaves as a solid at shear rates very close to zero. The yield stress for blood is very small, approximately in the range from 0.005 to 0.01 N/m^2.

Kinematic viscosity is another fluid property that has been used to characterize flow. It is the ratio of absolute viscosity to fluid density and

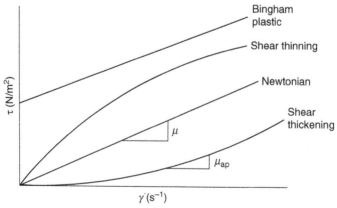

Figure 1.8 Shear stress versus rate of shearing strain for some non-Newtonian fluids.

is represented by the Greek character ν (nu). Kinematic viscosity can be defined by the equation:

$$\nu = \frac{\mu}{\rho}$$

where μ is the absolute viscosity and ρ is the fluid density.

The SI units for absolute viscosity are Ns/m^2. The SI units for kinematic viscosity are m^2/s.

1.2.5 Clinical feature: polycythemia

Polycythemia refers to a condition in which there is an increase in hemoglobin above 17.5 g/dL in adult males or above 15.5 g/dL in females (Hoffbrand and Pettit, 1984). There is usually an icrease in the number of red blood cells above $6 \times 10^{12}\ L^{-1}$ in males and $5.5 \times 10^{12}\ L^{-1}$ in females. That is, a sufferer from this condition has a much higher blood viscosity due to this elevated red blood cell count.

Symptoms of polycythemia are typically related to an increase in blood viscosity and clotting. The symptoms include headache, dizziness, itchiness, shortness of breath, enlarged spleen, and redness in the face.

Polycythemia vera is an acquired disorder of the bone marrow that results in an increase in the number of blood cells resulting from excessive production of all three blood cell types: erythrocytes, or red blood cells; leukocytes, or white blood cells; and thrombocytes, or platelets.

The cause of polycythemia vera is not well-known. It rarely occurs in patients under 40 years. Polycythemia usually develops slowly, and a patient might not experience any problems related to the disease even after being diagnosed. In some cases, however, the abnormal bone marrow cells grow uncontrollably resulting in a type of leukemia.

In patients with polycythemia vera, there is also an increased tendency to form blood clots that can result in strokes or heart attacks. Some patients may experience abnormal bleeding because their platelets are abnormal.

The objective of the treatment is to reduce the high blood viscosity (thickness of the blood) due to the increased red blood cell mass, and to prevent hemorrhage and thrombosis.

Phlebotomy is one method used to reduce the high blood viscosity. In phlebotomy, one unit (pint) of blood is removed, weekly, until the hematocrit is less than 45; then, phlebotomy is continued as necessary. Occasionally, chemotherapy may be given to suppress the bone marrow. Other agents such as interferon may be given to lower the blood count.

A condition similar to polycythemia may be experienced by high-altitude mountaineers. Due to a combination of dehydration and excess red blood cell production brought about by extended stays at high altitudes, the climber's blood thickens dangerously. A good description of what this is like is found in the later chapters of *Annapurna*, by Maurice Herzog.

1.3   Fundamental Method for Measuring Viscosity

A fundamental method for measuring viscosity involves a viscometer made from concentric cylinders. See Fig. 1.9. The fluid for which the viscosity is to be measured is placed between the two cylinders. The torque generated on the inner fixed cylinder by the outer rotating cylinder is determined by using a torque-measuring shaft. The force required to cause the cylinder to spin and the velocity at which it spins are also measured. Then the viscosity may be calculated in the following way:

The shear stress τ in the fluid is equal to the force F applied to the outer cylinder divided by the surface area A of the internal cylinder, that is,

$$\tau = \frac{F}{A}$$

The shear rate $\dot{\gamma}$ for the fluid in the gap, between the cylinders, may also be calculated from the velocity of the cylinder, V, and the gap width h as

$$\dot{\gamma} = \frac{V}{h}$$

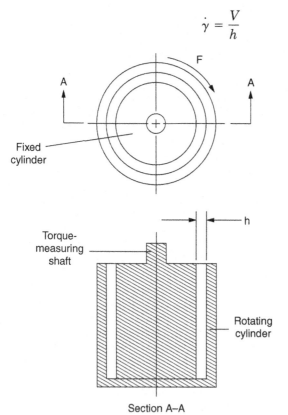

Figure 1.9a Cross section of a rotating cylinder viscometer.

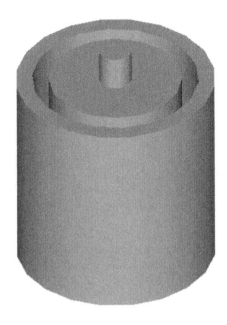

Figure 1.9b Rotating cylinder viscometer.

From the shear stress and the shear rate, the viscosity and/or the kinematic velocity may be obtained as

$$\mu = \frac{\tau}{\dot{\gamma}} \quad \text{and} \quad \nu = \frac{\mu}{\rho}$$

where μ = viscosity
ν = kinematic viscosity
ρ = density

A typical value for blood viscosity in humans is 0.0035 Ns/m^2, or 0.035 poise (P), or 3.5 cP. Note that 1 P = 1 dyne s/cm^2, or 0.1 Ns/m^2. Another useful pressure unit conversion is that 1 mmHg = 133.3 N/m^2.

Let T represent the measured torque in the viscometer shaft, and ω is its angular velocity in rad/s. Assume that D is the radius of the inner viscosimeter cylinder, and L is its length. The fluid velocity at the inner surface is

$$V = \omega \frac{D}{2}$$

It can be shown that

$$T = F \frac{D}{2}$$

leading to an equation which relates the torque, the angular velocity, and the geometric parameters of the device.

$$\mu = \frac{4T\,h}{\pi\,D^3\,L\,\omega}$$

1.3.1 Example problem: viscosity measurement

Whole blood (assume $\mu = 0.0035\ \text{Ns/m}^2$) is placed in a concentric cylinder viscometer. The gap width is 1 mm and the inner cylinder radius is 30 mm. Estimate the wall shear stress in the fluid. Assume the angular velocity of the outer cylinder to be 60 rpm.

We can begin by calculating the shear rate based on the angular velocity of the cylinder, its radius, and the gap between the inner and outer cylinders. The shear rate is equal to the velocity of the outer cylinder multiplied by the gap between the cylinders (see Fig. 1.9a). That is,

$$\dot{\gamma} = \frac{V}{h}$$

The wall shear stress is equal to the viscosity multiplied by the shear rate. Thus,

$$\tau = \mu\dot{\gamma} = \frac{\mu(r\omega)}{h} = 0.0035\,\frac{\text{Ns}}{\text{m}^2}\,\frac{\left(\dfrac{31}{1000}\right)\text{m}(60)\left(\dfrac{\pi}{30}\right)\dfrac{\text{rad}}{\text{s}}}{(1/1000)\text{m}} = 0.682\,\frac{\text{N}}{\text{m}^2}$$

1.4 Introduction to Pipe Flow

An Eulerian description of flow is one in which a field concept has been used. Descriptions which make use of velocity fields and flow fields are Eulerian descriptions. Another type of flow description is a Lagrangian description. In the Lagrangian description, particles are tagged and the paths of those particles are followed. An Eulerian description of goose migration could involve you or me sitting on the shore of Lake Erie and counting the number of geese that fly over a predetermined length of shoreline in an hour. In a Lagrangian description of the same migration, we might capture and band a single goose with a radio transmitter, and study the path of the goose, including its position and velocity as a function of time.

Consider an Eulerian description of flow through a constant–cross section pipe, as shown in Fig. 1.10. In the figure, the fluid velocity is shown across the pipe cross section and at various points along the length of

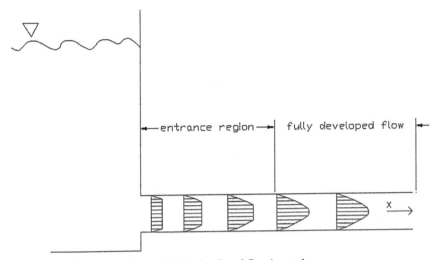

Figure 1.10   Entrance region and fully developed flow in a tube.

the pipe. In the entrance region, the flow begins with a relatively flat velocity profile and then develops an increasingly parabolic flow profile as the distance x along the pipe increases. Once the flow profile becomes constant and no longer changes with increasing x, the velocity profile does not change. The region in which the velocity profile is constant is known as the region of fully developed flow.

The pressure gradient is the derivative of pressure with respect to distance along the pipe. Mathematically, the pressure gradient is written as dP/dx, where P is the pressure inside the pipe at some point and x is the distance in the direction of flow. In the region of fully developed flow, the pressure gradient dP/dx is constant. On the other hand, the pressure gradient in the entrance region varies with the position x, as shown in the plot in Fig. 1.11. The slope of the plot is the pressure gradient.

1.4.1 Reynolds number

Osborne Reynolds was a British engineer, born in 1842 in Belfast. In 1895, he published the paper: "On the dynamical theory of incompressible viscous fluids and the determination of the criterion." The paper was a landmark contribution to the development of fluid mechanics and the crowning achievement in Reynolds' career. He was the first professor of engineering at the University of Manchester, England.

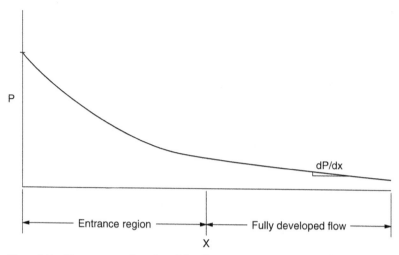

Figure 1.11   Pressure as a function of the distance along a pipe. Note that the pressure gradient *dP/dx* is constant for fully developed flow.

The Reynolds number is a dimensionless parameter named after Professor Reynolds. The number is defined as

$$\text{Re} = \frac{\rho VD}{\mu}$$

where ρ = fluid density in kg/m^3
  V = fluid velocity in m/s
  D = pipe diameter in m
  μ = fluid viscosity in Ns/m^2

Unless otherwise specified, this V will be considered to be the average velocity across the pipe cross section. Physically, the Reynolds number represents the ratio of inertial forces to viscous forces.

The Reynolds number helps us to predict the transition between laminar and turbulent flows. Laminar flow is highly organized flow along streamlines. As velocity increases, flow can become disorganized and chaotic with a random 3-D motion superimposed on the average flow velocity. This is known as turbulent flow. Laminar flow occurs in flow environments where Re < 2000. Turbulent flow is present in circumstances under which Re > 4000. The range of 2000 < Re < 4000 is known as the transition range.

The Reynolds number is also useful for predicting entrance length in pipe flow. I will denote the entrance length as X_E. The ratio of entrance length to pipe diameter for laminar pipe flow is given as

$$\frac{X_E}{D} \cong 0.06 \, \text{Re}$$

Consider the following example: If Re = 300, then X_E = 18 D, and an entrance length equal to 18 pipe diameters is required for fully developed flow. In the human cardiovascular system, it is not common to see fully developed flow in arteries. Typically, the vessels continually branch, with the distance between branches not often being greater than 18 pipe diameters.

Although most blood flow in humans is laminar, having a Re of 300 or less, it is possible for turbulence to occur at very high flow rates in the descending aorta, for example, in highly conditioned athletes. Turbulence is also common in pathological conditions such as heart murmurs and stenotic heart valves.

Stenotic comes from the Greek word "stenos," meaning narrow. Stenotic means narrowed, and a stenotic heart valve is one in which the narrowing of the valve is a result of the plaque formation on the valve.

1.4.2 Example problem: Reynolds number

Estimate the Reynolds number for blood flow in a retinal arteriole, using the published values from Gilmore et al. Assume that the blood density is 1060 kg/m^3. Is there any concern that blood flow in the human retina will become turbulent?

From Table 1.1, we see that the inside diameter of the arteriole is 0.008 cm, the mean velocity in the vessel is 3 cm/s, and the viscosity measured as 0.0035 Ns/m^2. The Reynolds number can be calculated as

$$\text{Re} = \frac{\rho V D}{\mu} = \frac{1060 \, \frac{\text{kg}}{\text{m}^3} \, \frac{3}{100} \, \frac{\text{m}}{\text{s}} \, \frac{0.008}{100} \, \text{m}}{0.0035 \, \frac{\text{Ns}}{\text{m}^2}} = 0.73$$

For this flow condition, the Reynolds number is far, far less than 2000, and there is no danger of the flow becoming turbulent.

1.4.3 Poiseuille's law

Velocity as a function of radius. In 1838, Jean-Marie Poiseuille empirically derived this law, which is also known as the Hagen-Poiseuille law, due to the additional experimental contribution by Gotthilf Heinrich Ludwig Hagen in 1839. This law describes steady, laminar, incompressible, and viscous flow of a Newtonian fluid in a rigid, cylindrical tube of constant cross section. The law was published by Poiseuille in 1840.

Consider a cylinder of fluid in a region of fully developed flow. See Fig. 1.12.

Then, I draw a free-body diagram of a cylinder of fluid and sum the forces acting on that cylinder. See Fig. 1.13 for a free-body diagram.

Next, we will need to make some assumptions. First, assume that the flow is steady. This means that the flow is not changing with time; that is, the derivative of flow rate with respect to time is equal to zero. Therefore,

$$\frac{dQ}{dt} = 0$$

Second, assume that the flow is through a long tube with a constant cross section. This type of flow is known as uniform flow. For steady flows in long tubes with a constant cross section, the flow is fully developed and, therefore, the pressure gradient dP/dx is constant.

Third, assume that the flow is Newtonian. Newtonian flow is flow in which the wall shearing stress τ in the fluid is constant. In other words, the viscosity does not depend on the shear rate $\dot{\gamma}$, and the whole process is carried out at constant temperature.

Now, let the x direction be the axial direction of the pipe with downstream (to the right) being positive. If the flow is not changing with time, then the sum of the forces in the x direction is zero, and Eq. (1.1) is written as

$$P(\pi\, r^2) - (P + dP)(\pi\, r^2) - \tau 2\pi\, r dx = 0 \tag{1.1}$$

Simplifying Eq. (1.1), we can write

$$-dP(\pi r^2) = 2\pi r \tau\, dx \tag{1.2}$$

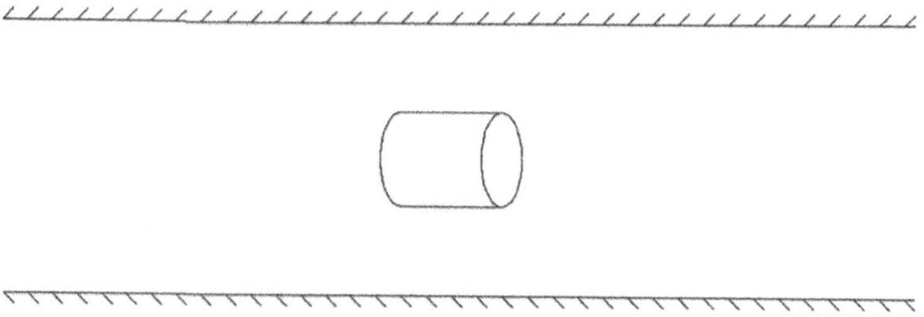

Figure 1.12   An element of fluid in pipe flow.

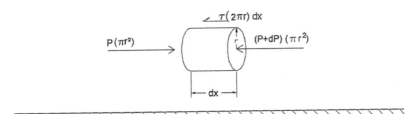

Figure 1.13   Free-body diagram of the forces acting on an element of fluid in pipe flow.

A simple force balance results in the shear stress as a function of the pressure gradient dP/dx and the radial position r as follows:

$$\tau = -r/2 \; dP/dx \tag{1.3}$$

and

$$\tau_{\text{wall}} = (-R_{\text{tube}}/2)(dP/dx) \tag{1.4}$$

Recall, from the definition of viscosity, that the shear stress is also related to the shear rate in the following way:

$$\tau = -\mu \; dV/dr \tag{1.5}$$

Now, if we solve Eqs. (1.4) and (1.5) together, the resulting expression relates the shear rate to the pressure gradient as

$$-r/2 \; dP/dx = -\mu \; dV/dr \tag{1.6}$$

Then, separating the variables, this expression becomes a differential equation with the variables, velocity V and radius r, as in the following:

$$dV = \frac{1}{2\mu} \frac{dP}{dx} r \, dr \tag{1.7}$$

The next step in the analysis is to solve the differential equation. The solution is obtained as

$$V = \frac{1}{2\mu} \frac{dP}{dx} \frac{r^2}{2} + C_1 \tag{1.8}$$

which gives the velocity of each point in the tube as a function of the radius r.

So far, in this analysis, we have made three assumptions: first, steady flow (i.e., $dQ/dt = 0$); second, fully developed flow in a constant–cross section tube (i.e., dP/dx is constant); and third, viscosity μ is constant. Now, we need to make the fourth assumption, which is the no-slip condition. This means that V at the wall is zero when r equals the radius of the tube.

Therefore, set $r = R_{tube} = R$ and $V = 0$ to solve for C_1. From Eq. (1.8), we can write

$$0 = \frac{1}{2\mu} \frac{dP}{dx} \frac{R^2}{2} + C_1 \tag{1.9}$$

or

$$C_1 = -\frac{1}{2\mu} \frac{dP}{dx} \frac{R^2}{2} \tag{1.10}$$

Then, Eq. (1.8), which gives the velocity as a function of the radius r, becomes

$$V = \frac{1}{4\mu} \frac{dP}{dx} \left[r^2 - R^2 \right] \tag{1.11}$$

The fifth and final assumption for this development is that the flow is laminar and not turbulent. Otherwise, this parabolic velocity profile would not be a good representation of the velocity profile across the cross section.

Note that in Eq. (1.11), dP/dx must have a negative value; as pressure drops, that would give a positive velocity. This is consistent with the definition of positive x as a value to the right, or downstream. Note also that the maximum velocity will occur on the arterial centerline. By substituting $r = 0$ into Eq. (1.11), the maximum velocity can be obtained and may be written as

$$V_{max} = \frac{1}{4\mu} \frac{dP}{dx} \left[-R^2 \right] \tag{1.12}$$

Furthermore, velocity as a function of radius may be written in a more convenient form:

$$V = V_{max} \left[1 - \frac{r^2}{R^2} \right] \tag{1.13}$$

1.4.4 Flow rate

Now that the flow profile is known and a function giving velocity as a function of radius can be written, it is possible to integrate the velocity multiplied by the differential area to find the total flow passing through the tube. Consider a ring of fluid at some distance r from the centerline of the tube. See Fig. 1.14.

The differential flow dQ passing through this ring may be designated as

$$dQ = 2\pi V r\, dr \qquad (1.14)$$

To obtain the total flow passing through the cross section, integrate the differential flow in the limits from $r = 0$ to $r = R_{\text{tube}}$, after substituting Eq. (1.11) for V. Thus,

$$Q = 2\pi \int_0^R V r\, dr \qquad (1.15)$$

$$Q = \frac{\pi}{2\mu}\frac{dP}{dx}\int_0^R (r^3 - rR^2)dr = \frac{\pi}{2\mu}\frac{dP}{dx}\left.\left(\frac{r^4}{4} - \frac{r^2 R^2}{2}\right)\right|_0^R \qquad (1.16)$$

This yields the expression:

$$Q = \frac{-\pi R^4}{8\mu}\frac{dP}{dx} \xrightarrow{\text{i.e.,}} \text{Poiseuille's law!} \qquad (1.17)$$

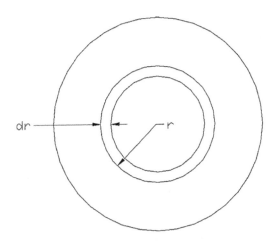

Figure 1.14 Cross section of a tube, showing a ring of fluid dr thick, at a radius of r.

Finally, it is now possible to solve for the average velocity across the cross section and to check assumption 5, that is, the flow is laminar. The expression for the average velocity across the cross section is

$$V_{avg} = \frac{Q}{A} = \frac{-R^2}{8\mu} \frac{dP}{dx} = \frac{V_{max}}{2} \tag{1.18}$$

To verify assumption 5, use V_{avg} to calculate the Reynolds number and check, to be sure, it is less than 2000. If the Reynolds number is greater than 2000, the flow may be turbulent, and Poiseuille's law no longer applies.

Remember to check whether $Re = \frac{\rho VD}{\mu}$ is less than 2000.

1.5   Bernoulli Equation

For some flows, it is possible to neglect the viscosity and focus on the pressure changes along a streamline, which is a line drawn through the flow field in a direction that is always tangential to the velocity field. For such an inviscid, incompressible flow along a streamline, the Bernoulli equation can be used to investigate the relationship between pressures and velocities. This is particularly useful in situations with converging flows. The equation, named after the Swiss mathematician, physicist, and physician, Daniel Bernoulli, is written as

$$P_1 + \frac{1}{2}\rho V_1^2 + \rho g z_1 = P_2 + \frac{1}{2}\rho V_2^2 + \rho g z_2$$

where P_1 and P_2 = pressures at points 1 and 2
V_1 and V_2 = velocities at points 1 and 2
z_1 and z_2 = heights at points 1 and 2

Because the Bernoulli equation does not take frictional losses into account, it is not appropriate to apply the Bernoulli equation to the flow through long constant–cross section pipes, as described by Poiseuille's law.

1.6   Conservation of Mass

If we want to measure mass flow rate in a very simple way, we might cut open an artery and collect the blood in a beaker for a specific interval of time. After weighing the blood in the beaker to determine the mass m, we would divide by the time t, required to collect the blood, to get the flow rate $Q = m/t$. However, this is not always a practical method of measuring the flow rate.

Instead, consider that, for any control volume, conservation of mass guarantees that the mass entering the control volume in a specific time is equal to the sum of the mass leaving that same control volume over the same time interval plus the increase in mass inside the control volume, that is,

$$\left(\frac{\text{Mass}}{\text{Time}}\right)_{\text{in}} \Delta t = \left(\frac{\text{mass}}{\text{time}}\right)_{\text{out}} \Delta t + (\text{mass increase})$$

The continuity equation presents conservation of mass in a slightly different form for cases where the mass inside the control volume does not increase or decrease, and it is written as

$$\rho_1 A_1 V_1 = \rho_2 A_2 V_2 = \text{constant}$$

where ρ_1 = the fluid density at point 1
A_1 = the area across which the fluid enters the control volume
V_1 = the average velocity of the fluid across A_1

The variables ρ_2, A_2, and V_2 are density, area, and average velocity at point 2.

For incompressible flows under circumstances where the mass inside the control volume does not increase or decrease, the density of the fluid is constant (i.e., $\rho_1 = \rho_2$) and the continuity equation can be written in its more usual form:

$$A_1 V_1 = A_2 V_2 = Q = \text{constant}$$

where Q is the volume flow rate going into and out of the control volume.

Consider the flow at a bifurcation, or a branching point, in an artery. See Fig. 1.15. If we apply the continuity equation, the resulting expression is

$$Q_1 = Q_2 + Q_3$$

or

$$A_1 V_1 = A_2 V_2 + A_3 V_3$$

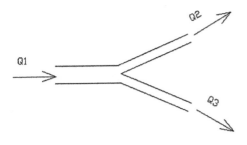

Figure 1.15   A bifurcation in an artery.

1.6.1 Venturi meter example

A Venturi meter measures the velocity of the flow using the pressure drop between two points on either side of a Venturi throat, as shown in Fig. 1.16.

First, suppose that the viscosity of the fluid were zero. We can apply the Bernoulli equation between points 1 and 2, as well as the continuity equation. Assume that the Venturi meter is horizontal and that $z_1 = z_2 = 0$.

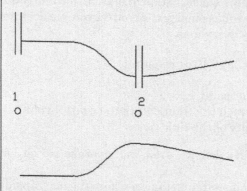

Figure 1.16 A Venturi flowmeter with pressure measuring ports at points 1 and 2.

(a) The Bernoulli equation is

$$P_1 + \frac{1}{2}\rho V_1^2 + gz_1 = P_2 + \frac{1}{2}\rho V_2^2 + gz_2$$

(b) The continuity equation is

$$A_1 V_1 = A_2 V_2 = Q \quad \text{or} \quad V_2 = \frac{A_1}{A_2}V_1$$

(c) Combining the equations in (a) and (b), and using the assumption, we can write

$$P_1 - P_2 = \frac{1}{2}\rho V_1^2\left[\left(\frac{A_1}{A_2}\right)^2 - 1\right]$$

(d) Therefore, solving for V_1, we obtain

$$V_1 = \sqrt{\frac{2(P_1 - P_2)}{\rho\left[\left(\frac{A_1}{A_2}\right)^2 - 1\right]}}$$

(e) Once we have solved for V_1, the ideal flow can be calculated by applying the continuity equation as

$$Q_{\text{ideal}} = V_1 A_1$$

(f) In practice, the flow is not inviscid. Because of frictional losses through the Venturi, it will be necessary to calibrate the flowmeter and adjust the actual flow using the calibration constant c. The actual flow rate is expressed as

$$Q_{actual} = cV_1A_1$$

1.7   Fluid Statics

In general, fluids exert both normal and shearing forces. This section reviews a class of problems in which the fluid is at rest. A velocity gradient is necessary for the development of a shearing force. So, in the case where acceleration is equal to zero, only normal forces occur. These normal forces are also known as hydrostatic forces.

In Fig. 1.17, a point P_1 in a fluid is shown at a depth of h below the surface of the fluid. The pressure exerted at a point in the fluid by the column of fluid above the point is

$$P_1 = \gamma h$$

where P_1 = the pressure at point 1
γ = the specific weight of the fluid
h = the distance between the fluid surface and the point P_1

or $$P_1 = \rho g h$$

where ρ = the fluid density
g = the acceleration due to gravity
h = the distance between the fluid surface and the point P_1

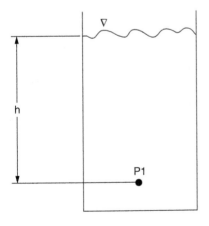

P1

Figure 1.17   Fluid in a reservoir showing the depth of point P_1.

1.7.1   Example problem: fluid statics

A liquid (viscosity = 0.002 Ns/m²; density = 1000 kg/m³) is pumped through the circular tube, as shown in Fig. 1.18. A differential manometer is connected to the tube to measure the pressure drop along the tube. When the differential reading h is 6 mm, what is the pressure difference between the points 1 and 2?

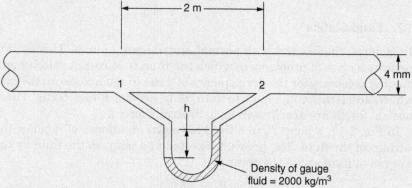

Figure 1.18   Pipeline using a differential manometer.

Solution: Assume that the pressure at point 1 is P_1. See Fig. 1.19. The pressure at point a is increased by the column of water between the points 1 and a. The specific gravity γ_{water} of the water is 9810 N/m³. If the vertical distance between 1 and a is d_a, then the pressure at a is

$$P_a = P_1 + \gamma_{\text{water}} \, d_a$$

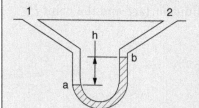

Figure 1.19   Schematic of the manometer.

The pressure at b can now be calculated from the pressure at a. Point b is at a higher position than point a. Therefore, the pressure at b is less than the pressure at a. The pressure at b is

$$P_b = P_a - \gamma_{\text{gauge}} \, h$$

$$= P_1 + \gamma_{\text{water}} \, d_a - \gamma_{\text{gauge}} \, h$$

where γ_{gauge} is the specific weight of the gauge fluid and h is the height of point b with respect to point a.

The pressure at 2 is also lower than the pressure at b and is calculated from

$$P_2 = P_b - (d_a - h)\gamma_{\text{water}}$$

$$= P_1 + \gamma_{\text{water}}\, d_a - \gamma_{\text{gauge}}\, h - (d_a - h)\gamma_{\text{water}}$$

Finally, the pressure difference between 1 and 2 is written as

$$P_1 - P_2 = (\gamma_{\text{gauge}} - \gamma_{\text{water}})h$$

or

$$\Delta P = \left(2000(9.81)\,\frac{N}{m^3} - 1000(9.81)\,\frac{N}{m^3}\right)\frac{6}{1000}\,m = 58.9\,\frac{N}{m^2}$$

1.8   The Womersley Number α: a Frequency Parameter for Pulsatile Flow

The Womersley number, or alpha parameter, is another dimensionless parameter that has been used in the study of fluid mechanics. This parameter represents a ratio of transient to viscous forces, just as the Reynolds number represented a ratio of inertial to viscous forces. A characteristic frequency represents the time dependence of the parameter. The Womersley number may be written as

$$\alpha = r\sqrt{\frac{\omega}{\nu}} \text{ or } \alpha = r\sqrt{\frac{\omega\rho}{\mu}}$$

where r = the vessel radius

ω = the fundamental frequency[7]

ρ = the density of the fluid

μ = the viscosity of the fluid

ν = the kinematic viscosity

In higher-frequency flows, the flow profile is blunter near the centerline of the vessel since the inertia becomes more important than viscous forces. Near the wall, where V is close to zero, viscous forces are still important.

---

[7] The fundamental frequency is typically the heart rate. The units must be rad/s for dimensional consistency.

Some typical values of the parameter α for various species are

human aorta	$\alpha = 20$
canine aorta	$\alpha = 14$
feline aorta	$\alpha = 8$
rat aorta	$\alpha = 3$

1.8.1 Example problem: Womersley number

The heart rate of a 400-kg horse is approximately 36 beats per minute (bpm), while the heart rate of a 3-kg rabbit is approximately 210 bpm (Li, 1996). Compare the Womersley number in the horse aorta to that in the rabbit aorta.

$$\alpha = r\sqrt{\frac{\omega}{\nu}} \ \text{ or } \ \alpha = r\sqrt{\frac{\omega\rho}{\mu}}$$

Li also points out that the size of red blood cells from various mammals does not vary much in diameter. No mammal has a red blood cell larger than 10 mm in diameter. Windberger et al. also published the values of viscosity for nine different animals, including horse and rabbit. For this example problem, we will assume that the blood density is approximately the same across the species. The published values of viscosity were 0.0052 Ns/m² for the horse and 0.0040 Ns/m² for the rabbit.

Allometric studies of mammals show that various characteristics, like aortic diameter, grow in a certain relationship with the size of the animal. From Li, we can use the published allometric relationship for the aorta size of various mammals to estimate the diameter of the aorta. Thus, we have (Li, 1996, Equation 3.7, p. 27)

$$D = 0.48 \, W^{0.34}$$

where D is the aortic diameter, given in centimeters, and W is the animal weight, given in kilograms. Therefore, for this problem, we will use an aortic diameter of approximately 3.7 cm for the horse, 2.0 cm for the man, and 0.7 cm for the rabbit. Then,

$$\alpha_{\text{horse}} = r\sqrt{\frac{\omega\rho}{\mu}} = \frac{3.7}{100}\,\text{m}\sqrt{\frac{36\left(\frac{\pi}{30}\right)\text{rad/s } 1060 \text{ kg/m}^3}{0.0052 \text{ Ns/m}^2}} = 32$$

$$\alpha_{\text{rabbit}} = r\sqrt{\frac{\omega\rho}{\mu}} = \frac{0.7}{100}\,\text{m}\sqrt{\frac{210\left(\frac{\pi}{30}\right)\text{rad/s } 1060 \text{ kg/m}^3}{0.004 \text{ Ns/m}^2}} = 17$$

From the results, one can see that transient forces become relatively more important than viscous forces as the animal size increases.

Review Problems

1. Show that for the Poiseuille flow in a tube of radius R, the wall shearing stress can be obtained from the relationship:

$$\tau_w = \frac{4\mu}{\pi R^3} Q$$

for a Newtonian fluid of viscosity μ. The volume rate of flow is Q.

2. Determine the wall shearing stress for a fluid, having a viscosity of 3.5 cP, flowing with an average velocity of 9 cm/s in a 3-mm-diameter tube. What is the corresponding Reynolds number? (The fluid density $\rho = 1.06 \text{ g/cm}^3$.)

3. In a 5-mm-diameter vessel, what is the value of the flow rate that causes a wall shear stress of 0.84 N/m²? Would the corresponding flow be laminar or turbulent?

4. It has been suggested that a power law:

$$\tau = -b \left(\frac{dv}{dr}\right)^n$$

be used to characterize the relationship between the shear stress and the velocity gradient for blood. The quantity b is a constant, and the exponent n is an odd integer. Use this relationship, and derive the corresponding velocity distribution for flow in a tube. Use all assumptions made in deriving Poiseuille's law (except for the shear stress relationship).
    Plot several velocity distributions (v/v_{max} versus r/R) for $n = 1, 3, 5, \ldots$, to show how the profile changes with n.

5. A liquid (viscosity $= 0.004 \text{ Ns/m}^2$; density $= 1050 \text{ kg/m}^3$) is pumped through the circular tube, as shown in Fig. A. A differential manometer is connected to the tube, as shown, to measure the pressure drop along the tube. When the differential reading h is 7 mm, what is the mean velocity in the tube?

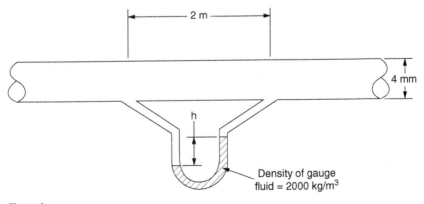

Figure A

6. A liquid (viscosity = 0.004 Ns/m^2; density = 1050 kg/m^3) is pumped through the circular tube, as shown in Fig. A. What is the height h, when the Reynolds number is equal to 1000?

7. The following data apply to the steady flow of blood through a long horizontal tube:

   tube diameter = 3 mm
   blood viscosity = 0.0035 Ns/m^2
   blood density = 1060 kg/m^3
   mean velocity = 4 cm/s
   (a) Is the flow laminar or turbulent?
   (b) Calculate, if possible, the shearing stress at the tube wall.
   (c) Calculate, if possible, the maximum velocity in the tube.

8. The following data apply to the steady flow of blood through a long horizontal tube:

   tube diameter = 3 mm
   blood viscosity = 0.0045 Ns/m^2
   blood density = 1060 kg/m^3
   mean velocity = 3 cm/s
   (a) Is the flow laminar or turbulent?
   (b) Calculate, if possible, the shearing stress at the tube wall.
   (c) Calculate, if possible, the maximum velocity in the tube.

9. The following data apply to the steady flow of blood through a long horizontal tube:

   tube diameter = 1 mm
   blood viscosity = 0.0030 Ns/m^2
   blood density = 1060 kg/m^3
   mean velocity = 8 cm/s
   (a) Is the flow laminar or turbulent?
   (b) Calculate, if possible, the shearing stress at the tube wall.
   (c) Calculate, if possible, the maximum velocity in the tube.

10. Arterial blood pressure is frequently measured by inserting a catheter into the artery. Estimate the maximum error (expressed in mmHg) that may exist in this measurement due to the possible creation of a stagnation point at the catheter tip. Assume the density of blood to be 1060 kg/m^3 and the peak velocity to be 110 cm/s. Make use of the steady-flow Bernoulli equation to obtain your answer.

11. A saline solution (density = 1050 kg/m^3) is ejected from a large syringe, through a small needle, at a steady velocity of 0.5 m/s. Estimate the pressure developed in the syringe. Neglect viscous effects. Assume that the velocity of the fluid in the large syringe is approximately zero, when compared with the velocity in the needle.

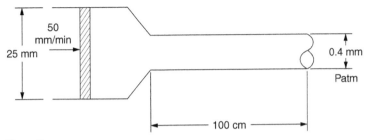

Figure B

12. A saline solution (density = 1050 kg/m^3) is ejected from a 20-mm-diameter syringe, through a 2-mm-diameter needle, at a steady velocity of 0.5 m/s. Estimate the pressure developed in the syringe. Neglect viscous effects.

13. Estimate the value of velocity at which the saline solution in problem 12 would be turbulent in the 2-mm-diameter needle. At that velocity, what is the pressure developed in the syringe?

14. A schematic of a 100-cm-long catheter having an inside diameter of 0.4 mm connected to a syringe is shown in Fig. B. In a typical infusion pump, the plunger is driven at a constant velocity, as illustrated. For a velocity of 50 mm/min, what volume rate of flow will discharge through the catheter, and what pressure will develop in the syringe? Neglect entrance effects, and assume the fluid to have a viscosity of 0.002 Ns/m^2 and a density of 1000 kg/m^3.

Bibliography

Acosta J de. *The Naturall and Morall Historie of the East and West Indies*, translated to English by Grimestone E. Valentine Simmes for Edward Blount and Williams Aspley booksellers, London; 1604.

Bacon, F. *Novum Organum*, edited by J. Devey. *New Four*, P.F. Collier, 1902 (English translation) pp. 356–360. http://www.constitution.org/bacon/nov.org.htm

Hoffbrand AV and Pettit JE. *Essential Haematology*. Blackwell Science Ltd., London; 1984.

John B. West. *High life; A History of High-Altitude Physiology and Medicine*. Oxford University Press, Oxford.

Milnor WR. *Cardiovascular Physiology*. Oxford University Press, New York; 1990: p. 334.

Gilmore ED, Hudson C, Preiss D, and Fisher J. Retinal arteriolar diameter, blood velocity and blood flow response to an isocapnic hyperoxic provocation, *Am J Physiol Heart Circ Physiol*, 2005; 288: H2912–H2917.

Li, JKJ, *Comparative Cardiovascular Dynamics of Mammals*. CRC Press, Boca Raton, FL; 1996.

Windberger U, Bartholovitsch A, Plasenzotti R, Korak KJ, and Heinze G. Whole blood viscosity, plasma viscosity and erythrocyte aggregation in nine mammalian species: Reference values and comparison data, *Exp Physiol*, 2003;88(3):431–440.

Cardiovascular Structure and Function

2.1 Introduction

One may question why it is important for a biomedical engineer to study physiology. To answer this question, we can begin by recognizing that cardiovascular disorders are now the leading cause of death in developed nations. Furthermore, to understand the pathologies or dysfunctions of the cardiovascular system, engineers must first begin to understand the physiology or proper functioning of that system. If an engineer would like to design devices and procedures to remedy those cardiovascular pathologies, then she or he must be well acquainted with physiology.

The cardiovascular system consists of the heart, arteries, veins, capillaries, and lymphatic vessels. Lymphatic vessels are vessels which collect extracellular fluid and return it to circulation. The most basic functions of the cardiovascular system are to deliver oxygen and nutrients, to remove waste, and to regulate temperature.

The heart is actually two pumps: the left heart and the right heart. The amount of blood coming from the heart (cardiac output) is dependent on arterial pressure. This chapter deals with the heart and its ability to generate arterial pressure in order to pump blood.

The adult human heart has a mass of approximately 300 g. If it beats 70 times per minute, then it will beat ~100,000 times per day, ~35 million times per year, and ~3 billion times (3×10^9) during your lifetime. If each beat ejects 70 mL of blood, your heart pumps over 7000 L, or the equivalent of 1800 gal/day. That is the same as 30 barrels of blood each and every day of your life! The lifetime equivalent work done by the heart is the equivalent of lifting a 30-ton weight to the top of Mount Everest. All of this work is done by a very hardworking 300-g muscle that does not rest!

2.2 Clinical Features

A 58-year-old German woman had undergone surgery in 1974 to fix a mitral stenosis.[1] She later developed a restenosis, which caused the mitral valve to leak. She underwent surgery in November 1982 to replace her natural mitral valve with a Björk-Shiley convexo-concave mitral prosthesis.[2]

In July 1992, she suffered sudden and severe difficulty breathing and was admitted to the hospital in her hometown in Germany. She had fluid in her lungs, and the diagnosis was that she had a serious hemodynamic disturbance due to inadequate heart function (cardiogenic shock). The physician listened for heart sounds but did not hear the normal "click" associated with a prosthetic valve. The patient's heart rate was 150 bpm, her systolic blood pressure was 50 mmHg, and her arterial oxygen saturation was 78 percent. Echocardiography[3] was performed, and no valve disk was seen inside the valve ring. An x-ray (see Fig. 2.1) appeared to show the valve disk in the abdominal aorta and the outflow strut in the pulmonary vein. Emergency surgery was performed, the disk and outflow strut were removed, and a new Medtronic-Hall mitral valve prosthesis was implanted into the woman. She eventually recovered.

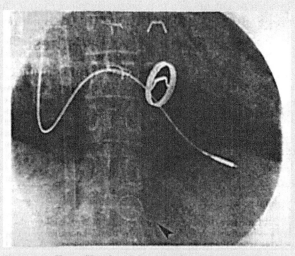

Figure 2.1 X-ray film showing escaped strut and disk (arrow). (*Courtesy of Herse et al., 1993, 41, p. 78, Figure 1.*)

---

[1]Narrowing or stricture.

[2]Artificial body part.

[3]Graphically recording the position and motion of the heart walls and internal structures.

2.3 Functional Anatomy

The cardiovascular system can be further divided into three subsystems. The **systemic circulation,** the **pulmonary circulation,** and the **coronary circulation** are the subsystems that, along with the heart and lungs, make up the cardiovascular system. See Fig. 2.2.

The three systems can be divided functionally based on the tissue to which they supply oxygenated blood. The systemic circulation is the subsystem supplied by the aorta that feeds the systemic capillaries. The pulmonary circulation is the subsystem supplied by the pulmonary artery that feeds the pulmonary capillaries. The coronary circulation is the specialized blood supply that perfuses cardiac muscle. (Perfuse means to push or pour a substance or fluid through a body part or tissue.)

The cardiovascular system has four basic functions:

1. It supplies oxygen to body tissues.

2. It supplies nutrients to those same tissues.

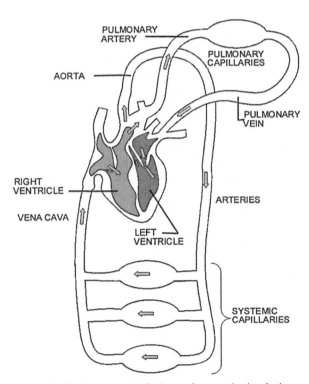

Figure 2.2 Pulmonary circulation and systemic circulation.

3. It removes carbon dioxide and other wastes from the body.

4. It regulates temperature.

The path of blood flowing through the circulatory system and the pressures of the blood at various points along the path tell us much about how tissue is perfused with oxygen. The left heart supplies oxygenated blood to the aorta at a relatively high pressure.

Blood continues to flow along the path through the circulatory system, and its path may be described as follows: Blood flows into smaller arteries and finally into systemic capillaries where oxygen is supplied to the surrounding tissues. At the same time, it picks up waste carbon dioxide from that same tissue and continues flowing into the veins. Eventually, the blood returns to the vena cava. From the vena cava, deoxygenated blood flows into the right heart. From the right heart, the still deoxygenated blood flows into the pulmonary artery. The pulmonary artery supplies blood to the lungs where carbon dioxide is exchanged with oxygen. The blood, which has been enriched with oxygen, flows from the lungs through the pulmonary veins and back to the left heart.

It is interesting to note that blood flowing through the pulmonary artery is deoxygenated and blood flowing through the pulmonary vein is oxygenated. Although systemic arteries carry oxygenated blood, it is a mistake to think of arteries only as vessels that carry oxygenated blood. A more appropriate distinction between arteries and veins is that arteries carry blood at a relatively higher pressure than the pressure within the corresponding veins.

2.4   The Heart as a Pump

The heart is a four-chambered pump and supplies the force that drives blood through the circulatory system. The four chambers can be broken down into two upper chambers, known as **atria,** and two lower chambers, known as **ventricles.** Check valves between the chambers ensure that the blood moves in only one direction and enables the pressure in the aorta, for example, to be much higher than the pressure in the lungs, restricting blood from flowing backward from the aorta toward the lungs.

The four chambers of the heart, namely, the right atrium, right ventricle, left atrium, and left ventricle, are shown in Fig. 2.3. Blood enters the right atrium from the vena cava. From the right atrium, blood is pumped into the right ventricle. From the right ventricle, blood is pumped downstream through the pulmonary artery to the lungs where it is enriched with oxygen and gives up carbon dioxide.

On the left side of the heart, oxygen-enriched blood enters the left atrium from the pulmonary vein. When the left atrium contracts, it

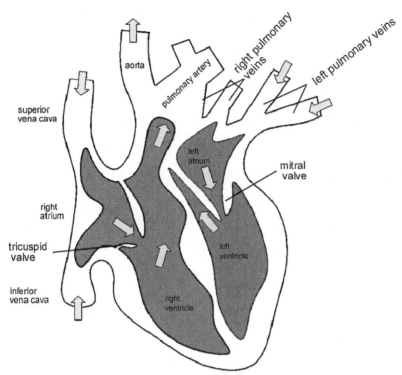

Figure 2.3   Heart chambers and flow through the heart. Direction of blood flow is indicated by blue arrows.

pumps blood into the left ventricle. When the left ventricle contracts, it pumps blood to a relatively high pressure, ejecting it from the left ventricle into the aorta.

2.5   Cardiac Muscle

The **myocardium,** or **muscle tissue of the heart,** is composed of millions of elongated, striated, multinucleated cardiac muscle cells. These cells are approximately 15 microns by 15 microns by 150 microns long and can be depolarized and repolarized like skeletal muscle cells. The meaning of the terms "depolarized" and "repolarized" will be explained in the next section. Figure 2.4 shows a typical group of **myocardial muscle cells.** Individual cardiac muscle cells are interconnected by dense structures known as **intercalated disks.** The cells form a latticework of muscular tissue known as a **syncytium.**

A multinucleated mass of cardiac muscle cells form a functional syncytium (pronounced sin–sish' e-um). The heart has two separate muscle

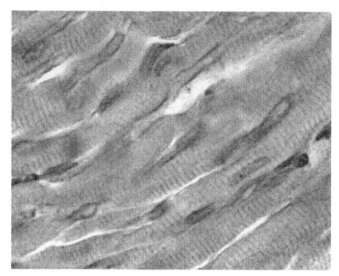

Figure 2.4 Photomicrograph of myocardial muscle. Available at
http://www.mhprofessional.com/product.php?isbn=0071472177

syncytia. The first is the muscle mass that makes up the two atria, and
the second is the muscle mass that makes up the two ventricles. The two
syncytia are separated by fibrous rings that surround the valves between
atria and ventricles. When one muscle mass is stimulated, the action
potential spreads over the entire syncytium. As a result, under normal
circumstances, both atria contract simultaneously and both ventricles
contract simultaneously.

Contraction in myocardium takes 10 to 15 times longer than it takes
in average skeletal muscle. Myocardium contracts more slowly because
sodium/calcium channels in myocardium are much slower than the
sodium channel in skeletal muscle, during repolarization. In addition,
immediately after the onset of the action potential, the permeability of
cardiac muscle membrane for potassium ions decreases about fivefold.
This effect does not happen in skeletal muscle. This decrease in permeability prevents a quick return of the action potential to its resting
level.

2.5.1 Biopotential in myocardium

Heart muscles operate due to electrical voltages which build up and
diminish in the individual muscle cells. The cellular membranes of
myocardial cells are electrically polarized in the resting state like any
other cells in the body. The resting, **transmural electrical potential
difference,** is approximately -90 mV in ventricular cells. The inside

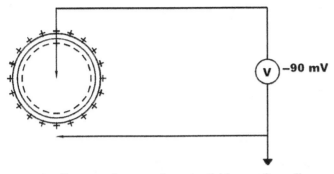

Figure 2.5   Transmembrane resting potential in a cardiac cell.

of the cell is negative with respect to the outside. This transmembrane potential exists because the cell membrane is selectively permeable to charged particles. Figure 2.5 shows the transmembrane resting potential in a cardiac cell.

The principal electrolyte ions inside myocardial cells, which are responsible for the transmembrane potential, are sodium, potassium, and chloride. Negative ions, or anions, associated with proteins and other large molecules are also very important for the membrane potential. They attract the positive potassium ion that can go inside the cell.

The permeability of the membrane to sodium ions is very low at rest, and the ions cannot easily pass through the membrane. On the other hand, permeability of the membrane with respect to both potassium and chloride ions is much higher, and those ions can pass relatively easily through the membrane. Since sodium ions cannot get inside the cell and the concentration of positively charged sodium ions is higher outside the cell, a net negative electrical potential results across the cell membrane.

The net negative charge inside the cell also causes potassium to concentrate inside the cell to counterbalance the transmembrane potential difference. However, the osmotic pressure caused by the high concentration of potassium prevents a total balancing of the electrical potential across the membrane.

Since some sodium continually leaks into the cell, maintaining a steady-state balance requires continual active transport. An active sodium-potassium pump in the cell membrane uses energy to pump sodium out of the cell and potassium into the cell.

2.5.2 Excitability

Excitability is the ability of a cell to respond to an external excitation. When the cell becomes excited, the membrane permeability changes,

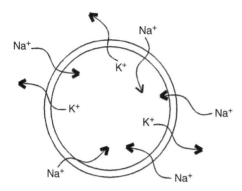

Figure 2.6 Depolarization of a myocardial cell.

allowing sodium to freely flow into the cell. In order to maintain equilibrium, potassium, which is at a higher concentration inside the cell, flows to the outside. In Fig. 2.6, a cardiac muscle cell is shown depolarizing.

In order to obtain a regulated **depolarization,** it is crucial that the increases in sodium and potassium permeabilities are offset in time. The sodium permeability must increase at the beginning of depolarization, and the potassium channel permeability increases during repolarization. It is also important that the potassium channels that open during an action potential are different from the leak channels that allow potassium to pass through the membrane at rest. In cardiac muscle, the action potential, or biopotential that results in muscle movement, is carried mainly by calcium from the extracellular space rather than by sodium. This calcium is then used to trigger the release of intracellular calcium to initiate contraction.

The ability of a cell to respond to excitation depends on the elapsed time since the last contraction of that cell. The heart will not respond to a new stimulation until it has recovered from the previous stimulation. That is, if you apply stimulation below the threshold before the non-responsive or refractory period has passed, the cell will not give a response. In Fig. 2.7, the **effective refractory period** (ERP) for a myocardial cell is shown to be approximately 200 ms. After the **relative refractory period** (RRP), the cell is able to respond to stimulation if the stimulation is large enough. The time for the relative refractory period in the myocardium is approximately another 50 ms.

During a short period following the refractory period, a period of **supernormality** (SNP) occurs. During the period of supernormality, the cell's transmembrane potential is slightly higher than its resting potential.

The refractory period in heart muscle is much longer than in skeletal muscle, because repolarization is much slower. Ventricular muscle in dogs has a refractory period of 250 to 300 ms at normal heart rates. The refractory period for mammalian skeletal muscle is 2 to 4 ms and 0.05 ms for mammalian nerve fiber.

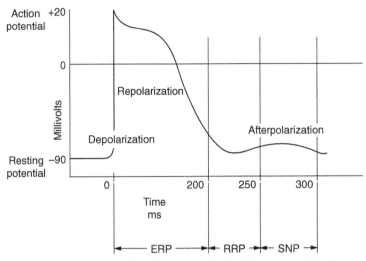

Figure 2.7 Action potential of a myocardial cell showing the effective refractory period (ERP), the relative refractory period (RRP), and the supernormality period (SNP).

2.5.3 Automaticity

Automaticity is the ability of a certain class of cells to depolarize spontaneously without external stimulation. All cardiac cells in the conduction system have automaticity. The **sinoatrial (SA) node** is a small group of cardiac muscle fibers on the posterior wall of the right atrium, and the muscle cells that make up the SA node have especially strong automaticity. Under normal circumstances, the cells of the SA node initiate the electrical activity of a heartbeat and control heart rate. They form the pacemaking site of the heart. The **atrioventricular (AV) node** is located at the lower interatrial septum. The AV node provides a limiting effect on the maximum heart rate. The AV node cannot transmit impulses faster than about 200 bpm. If the atrial rate is higher, as in atrial tachycardia, then some beats will be transmitted to the ventricle and others will not. This condition is known as atrioventricular block.

2.6 Electrocardiograms

The electrical activity of the heart is a useful indicator of how well the heart is functioning. Muscle cells possess an inherent electrical activity that is necessary to their function. Because the heart is a large mass of muscle cells, the resulting composite picture of the electrical activity of that mass is a clinically useful piece of information.

The composite electrical activity that results from the action of the total number of cardiac muscle fibers results in a signal that can be approximately represented by a vector quantity. Imagine for a moment, the electrical field that would be generated on the surface of the body as a result of a battery "spinning" in 3-D space inside your body. In this vector model, the heart is an electric dipole located in the thorax. The dipole can be represented by the **dipole moment vector** $\vec{M}$. $\vec{M}$ points from the negative to the positive "pole," or charge, and has a magnitude equal to the charge multiplied by the separation distance.

By placing electrodes on the appropriate location on the body surface, an electrocardiogram (ECG) can be measured. See Fig. 2.8. We label the standard characteristic points on the ECG using the letters P through U, where the points represent the following heart muscle activities:

- P—depolarization of the atria
- QRS—depolarization of the ventricles
- T—ventricular repolarization
- U—inconstant, when present it is believed to be due to after potentials

The magnitude and direction of $\vec{M}$ varies throughout the cardiac cycle. The vector is longer during ventricular depolarization. We can measure that magnitude and direction with pairs of electrodes.

2.6.1 Electrocardiogram leads

A pair of electrodes makes up a **lead.** The lead vector is a unit vector defining the direction of a given electrode pair. When $\vec{M}$ is parallel to a given lead, or electrode pair, then that lead will show the maximum output. In Fig. 2.9, the bipolar moment $\vec{M}$ has the component ν_{a1} along lead I. The direction of lead I is shown by the unit vector $\vec{a}_1$.

Figure 2.10 shows the directions for leads I, II, and III. Note that the direction for lead I is horizontal from the right arm toward the left arm. The dipole moment is positive in the direction toward the left arm, with respect to the right arm electrode, when the lead I vector is positive.

The direction of lead II is from right arm to left leg, and the direction for lead III is from left arm to left leg. **Einthoven's triangle** is an equilateral triangle whose legs are defined by the direction of the three leads. See Fig. 2.11. With the dipole moment plotted in the middle of

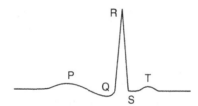

Figure 2.8 A typical wave from a single heartbeat of an electrocardiogram.

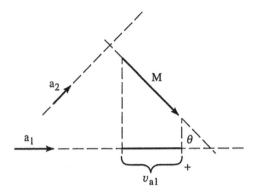

Figure 2.9 The component of $\vec{M}$ along lead I is ν_{a1}. The unit vector along lead I is $\vec{a}_1$.

Einthoven's triangle, the triangle can be used to show the components along each lead. Conversely, if the components are known from an ECG, the direction of the dipole moment vector can be calculated.

2.6.2 Mean electrical axis

The mean electrical axis (MEA) defines the direction of the dipole moment in the heart at the time of left ventricular systole. That is, the MEA is defined at the instant of the QRS complex. At that instant in

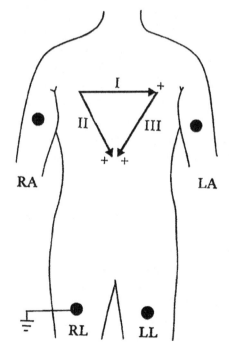

Figure 2.10 Direction of lead I, II, and III of the standard electrocardiogram. (*Modified from Webster, 1998, p. 237, Figure 6.3.*)

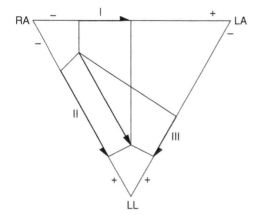

Figure 2.11 Einthoven's triangle for the standard electrocardiogram.

time, one can measure the magnitude of the components of two leads, and from this information, using Einthoven's triangle, it is possible to calculate the direction of the dipole moment.

The MEA is the angle between lead I and the dipole moment vector, as shown in Fig. 2.12. The MEA is also known as the QRS axis. The MEA depicts the heart's "electrical position" in the frontal plane during ventricular depolarization.

The MEA is normally between −30° and +120° with positive being clockwise from lead I. A pathologic left axis is between −30° and −90°, and a pathologic right axis is between 120° and 180°. There is a zone of uncertainty, in which the MEA is between −90° and −180°, that can indicate either an extreme right or extreme left pathologic axis.

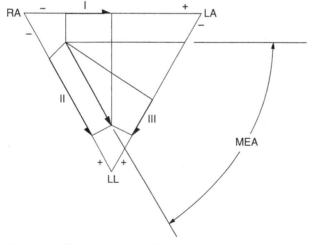

Figure 2.12 Dipole moment and the three components it generates along the three electrocardiogram axes.

2.6.3   Example problem: mean electrical axis

Given the ECG tracings from lead I and lead II from a single patient, shown in Fig. 2.13, find this patient's mean electrical axis (MEA). The magnitude of the QRS complex on lead I at the time of ventricular systole is 5 mV, and the magnitude of the QRS complex on lead II is 6 mV. Is the MEA in the normal range?

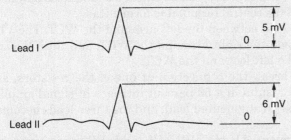

Figure 2.13   Electrocardiogram leads I and II from a patient for whom we want to measure the mean electrical axis.

Begin this problem by setting the magnitude of M_I = 5 mV. Now, find the horizontal component of the dipole moment vector M_X. Since the unit vector along lead I is the horizontal, x-axis, the magnitude of the dipole moment vector is the magnitude of lead I, that is, 5 mV.

$$|M_I| = |\vec{M_x}\hat{i}| = 5 \text{ mV} \qquad M_X = 5 \text{ mV}$$

Now, setting the magnitude of M_{II} = 6 mV and solving for M_y, we obtain

$$M_{II} = 6 \text{ mV} = (5\hat{i} + M_y\hat{j}) \cdot \hat{u}_{II}$$

$$6 \text{ mV} = (5\hat{i} + M_y\hat{j}) \cdot (\cos(60)\hat{i} - \sin(60)\hat{j})$$

$$6 \text{ mV} = ((5)(0.5) \text{ mV} + (M_y)(-0.866))$$

$$6 \text{ mV} = 2.5 \text{ mV} - 0.866 M_y \qquad M_y = 4.04 \text{ mV}$$

The dipole moment vector at the time of the QRS complex is now

$$\vec{M} = (5\hat{i} - 4.04\hat{j}) \text{ mV}$$

The magnitude of the vector at that instant is

$$\vec{M} = \sqrt{5^2 + 4.04^2} = 6.43 \text{ mV}$$

The angle is measured from the horizontal axis (lead I) with the value in the clockwise direction being a positive angle. The direction of the dipole moment vector, that is, the MEA, is

$$MEA = \tan^{-1}\left(\frac{-4.04}{5}\right) = 38.9° = 40°$$

The dipole moment vector of this patient, at the time of ventricular depolarization, can also be written as

$$\vec{M} = 6.4 \text{ mV} \angle 40°$$

2.6.4   Unipolar versus bipolar
and augmented leads

The **Wilson central terminal** (WCT) is a single **reference electrode** (not a lead). It gives the average of three voltages. The WCT equivalent location is that of an electrode approximately at the center of the body. Using a WCT as one electrode, it is possible to create three more **unipolar leads** called VL, VR, and VF. These are known as unipolar leads because they use only one electrode and the central terminal to form a lead.

The VL lead is the lead between the left arm and the WCT. The VR lead is the lead between the right arm and the WCT, and the VF lead is the lead between the left foot and the WCT.

In practice, if you break the connection at one of the resistors, as shown in Fig. 2.14, it results in a 50 percent increase in signal amplitude. This is known as an augmented lead, and the three leads become aVR, aVL, and aVF.

Using the three augmented leads, aVR, aVL, and aVF, in combination with the standard leads I, II, and III, we now have a lead that is positioned every 30° in the frontal plane as can be seen in Fig. 2.15. Note that leads I and aVF give the vertical and horizontal axes.

An engineer will quickly note that any vector in the frontal plane can be represented by any two of the leads, say I and aVF. Sometimes, when the dipole moment is exactly perpendicular to a chosen lead, then the component of that moment along that particular lead is approximately zero. In that case, the chosen lead is probably not an appropriate lead for measuring that particular patient's dipole moment at that point in time.

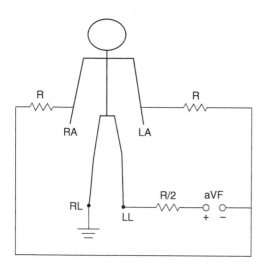

Figure 2.14 The circuit resulting in augmented lead aVF. Note that by breaking the connection between the left leg and the Wilson central terminal, the output of the aVF lead is increased by 50% over the VF lead. (*Modified from Webster, p. 240, Figure 6.5c.*)

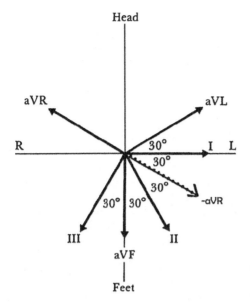

Figure 2.15 Directions from each of the standard bipolar and augmented unipolar leads. Note that these leads, taken in combination with the negative of each lead, are spaced every 30° in the frontal plane. (*Modified from Webster, p. 240, Figure 6.5d.*)

2.6.5 Electrocardiogram interpretations

Electrocardiograms are used to diagnose deviations of the MEA, heart arrhythmias, and heart blocks. A **heart arrhythmia** is a condition resulting in an abnormal heart rhythm. A **sinus rhythm** results from the normal pacing of the heart by the SA node. The ECG shows the normal p, q, r, s, and t waves. It is possible, and sometimes normal, for the patient to have a sinus rhythm and for the heart to beat faster or slower than normal.

Sinus tachycardia is the condition resulting in a faster than normal heart rate. Tachycardia often occurs in normal hearts in response to disease or physiologic or pharmacologic stimulus. Exercise often results in tachycardia—an increased heart rate.

Sinus brachycardia is a slower than normal heart rate. Once again, this slow heart rate occurs in normal persons (especially athletes). Sinus brachycardia also results from pharmacologic agents and disease of the SA node.

Atrial tachycardia results from an ectopic focus in either atrium. An ectopic focus means that there is an "extra" location, in the atria but outside of the SA node, from which the electrical pacing of the atria begins. In atrial tachycardia, the atrium beats too quickly, but may not be related to the rate of the ventricles. This is a common arrhythmia, seen in normal healthy persons of all ages.

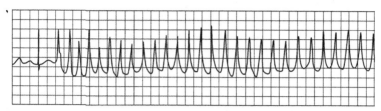

Figure 2.16   An electrocardiogram showing a case of the arrhythmia known as ventricular tachycardia. (*Modified from Selkurt, 1982, p. 287, Figure 8.13G.*)

Ventricular tachycardia is a serious condition that usually occurs in persons with heart disease. In ventricular tachycardia, the ventricle races at near maximal speed. An ECG showing ventricular tachycardia is shown in Fig. 2.16.

Ventricular fibrillation is characterized by a chaotic electrical activity of the heart. Because the electrical signal is chaotic and the muscle fibers do not depolarize in an organized way, no blood is pumped. This condition can be fatal within minutes if it remains uncorrected. Treatment of ventricular fibrillation is to impose more ventricular fibrillation. The entire heart is electrically depolarized with the hope that it will repolarize and begin beating synchronously, with the initiating electrical signal coming from the SA node.

2.6.6   Clinical feature: near maximal exercise stress test

A near maximal exercise stress test is sometimes used to evaluate heart function in a patient. An ECG is taken during exercise on a treadmill or stationary bicycle.

The patient tries to achieve 90 percent of his or her maximal heart rate based on the patient's age. For example, the maximal heart rate is 177 for an untrained 20 year old, 171 for a trained 20 year old, and 162 for a trained 45 year old. A physician should be on hand during the test, with a defibrillator available.

The PR interval of the ECG should be used as a baseline for ST interval displacement, with at least three consecutive complexes on a horizontal straight line drawn through the R point of the waves.

The ST displacement can be expressed in mm or in mV, with 1 mm being equal to 0.1 mV on a standard ECG. An ST displacement of 1 mm (0.1 mV) upward or downward indicates abnormal heart function. Figure 2.17 shows a normal heartbeat compared to a single beat in an ischemic heart. An ischemic heart is one for which the heart muscle tissue does not receive sufficient oxygen-carrying blood flow during maximum exercise.

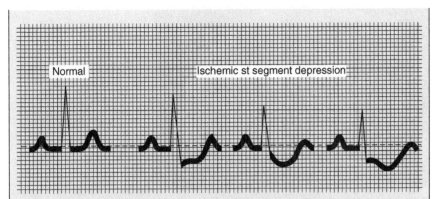

Figure 2.17 An electrocardiogram from an ischemic heart with an S-T segment depression and the diagnosis of agina pectoris. (*With permission from Gazes PC, Culler MR, and Stokes TC. The diagnosis of angina pectoris, Am Heart J, 1964; 67:830.*)

2.7 Heart Valves

Four cardiac valves help to direct flow through the heart. Heart valves cause blood to flow only in the desired direction. If a heart without these valves were to contract, it would compress the blood, causing it to flow both backward and forward (upstream and downstream). Instead, under normal physiological conditions, heart valves act as check valves to prevent blood from flowing in the reverse direction. In addition, heart valves remain closed until the pressure behind the valve is large enough to cause blood to move forward.

Each human heart has two atrioventricular (AV) valves which are located between the atria and the ventricles. The tricuspid valve is the valve between the right atrium and the right ventricle. The mitral valve is the valve between the left atrium and the left ventricle. The mitral valve prevents blood from flowing backward into the pulmonary veins and therefore into the lungs, even when the pressure in the left ventricle is very high. The mitral valve is a bicuspid valve, which has two cusps, while a tricuspid valve has three cusps.

The other two valves in the human heart are known as semilunar valves. The two semilunar valves are the aortic valve and the pulmonic valve. The aortic valve is located between the aorta and the left ventricle, and when it closes, it prevents blood from flowing backward from the aorta into the left ventricle. An aortic valve is shown in Fig. 2.18. The pulmonic valve is located between the right ventricle and the pulmonary artery, and when it closes, it prevents blood from flowing backward from the pulmonary artery into the right ventricle.

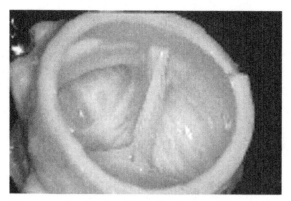

Figure 2.18   Aortic valve. (*From Lingappa and Farey, 2000.*)

Papillary muscles are cone-shaped projections of myocardial tissue that connect from the ventricle wall to the chordae tendineae, which are tendons connected to the edge of AV valves. The chordae tendineae are fine fibrous cords of collagen. These structures allow the AV valves to open and close, but they constrain the valves and prevent them from prolapsing, or collapsing backward, into the atria. Functionally, the papillary muscles and the chordae tendineae are part of the valve complex with which they are associated.

2.7.1   Clinical features

Chordae tendineae rupture and papillary muscle paralysis can be consequences of a heart attack. This can lead to bulging of the valve, excessive backward leakage into the atria (regurgitation), and even valve prolapse. Valve prolapse is the condition under which the valve inverts backward into the atrium. Because of these valve problems, the ventricle doesn't fill efficiently. Significant further damage, and even death, can occur within the first 24 h after a heart attack, because of this problem.

2.8   Cardiac Cycle

Figure 2.19 is a graphical representation of the cardiac cycle during 1 heartbeat. Figure 2.20 also shows typical pressure in the aorta, left ventricle, and left atrium as a function of time.

Systole is the term used to describe the portion of the heartbeat during which ejection takes place. Figure 2.19 shows atrial systole occurring over a period of 0.1 s, just before ventricular systole. Ventricular systole is shown to occur over a time period of 0.3 s, directly after atrial systole.

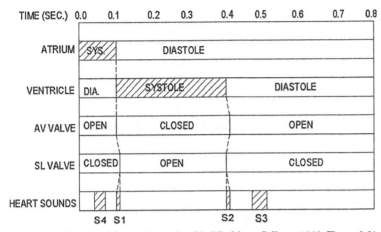

Figure 2.19 Timing of the cardiac cycle. (*Modified from Selkurt, 1982, Figure 8.3.*)

During atrial systole, pressure is generated in the atria, which forces the AV valves to open. The AV valves include the mitral and tricuspid valves. The AV valves remain open during atrial systole, until ventricular systole generates enough ventricular pressure to force those valves to close.

Diastole is the term used to describe the portion of the heartbeat during which chamber refilling is taking place. The ventricles undergo refilling, or ventricular diastole, during atrial systole. When the ventricle is filled and ventricular systole begins, the AV valves are closed and the atria undergo diastole. That is, the atria begin refilling with blood.

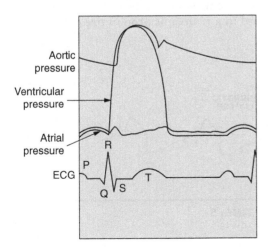

Figure 2.20 Typical pressures in the aorta, the left ventricle, and the left atrium are plotted against time.

Figure 2.19 shows a time period of 0.4 s after ventricular systole, when both the atria and the ventricles begin refilling and both heart chambers are in diastole. During the period when both heart chambers are refilling, or undergoing diastole, both AV valves are open and both the aortic and pulmonic valves are closed. The aortic and pulmonic valves are also called semilunar valves.

The semilunar valves are always closed except for a short period of ventricular systole when the pressure in the ventricle rises above the pressure in the aorta for the left ventricle and above the pressure in the pulmonary artery for the right ventricle. When the pressure in the left ventricle drops below the pressure level in the aorta, the pressure in the aorta forces the aortic valve to close to prevent blood from flowing backward into the left ventricle. Similarly, when the pressure in the right ventricle drops below the pressure levels in the pulmonary artery, the pressure in the pulmonary artery forces the pulmonic valve to close to prevent blood from flowing backward into the right ventricle.

During ventricular systole, the ventricular pressure waveform rises rapidly, and when the ventricular pressure rises above the aortic pressure, the aortic valve opens. The aortic valve remains open while the aortic pressure is lower than the left ventricular pressure. When the ventricular pressure drops below the aortic pressure, the aortic valve closes to prevent backflow into the ventricle from the aorta.

Figure 2.21 shows a more detailed aortic pressure-time curve. Systolic pressure is the largest pressure achieved during systole. Diastolic pressure is the lowest aortic pressure measured during diastole. The pulse pressure is defined as the difference between the systolic and diastolic pressures. The anacrotic limb of the pressure waveform is the ascending limb of the pressure waveform, and the descending portion of the waveform is known as the catacrotic limb. A notch exists in the normal catacrotic limb of the aortic pressure waveform. That notch is known as the dicrotic notch and is related to a transient drop in pressure due to the closing of the aortic valve.

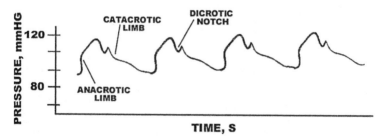

Figure 2.21 Aortic pressure waveform.

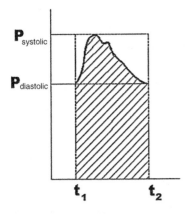

Figure 2.22 A single pressure waveform plotted against time

Figure 2.22 shows a close-up view of a single pulse from an aortic pressure-time curve. Systolic pressure, diastolic pressure, and pulse pressure are shown graphically in the figure. It is possible to find the mean pressure of a beat by integrating to find the area under the pressure curve, $P(t)$, and then dividing by the time interval between t_1 and t_2. In the graph shown, the mean pressure P_{mean} can be found from

$$P_{mean} = \frac{1}{t_2 - t_1} \int_{t_1}^{t_2} P(t)\,dt$$

2.8.1 Pressure-volume diagrams

Pressure in the heart is related to volume in the heart and vice versa. How the two parameters are related depends on things like the geometry of the heart, the passive material properties of the heart, and the active participation of muscle tissue in changing the shape of the heart. Because of these useful but somewhat complex relationships, **pressure-volume diagrams** provide a wealth of information about heart function at a given set of circumstances.

If the heart stopped beating completely, the pressure in the entire circulatory system would come to some equilibrium pressure. That equilibrium pressure would be greater than zero. What would cause that nonzero pressure in the absence of the heartbeat? The volume of blood in the circulatory system is held in place by the geometry of the vessels of the circulatory system. The stiffness of those vessels also adds something to that pressure. This very small, but nonzero, pressure is known as the **mean circulatory filling pressure.**

To continue, the **stroke volume (SV)** of the heart is defined as the **end-diastolic volume (EDV)** (ventricular volume at the end of the filling period) minus the **end-systolic volume (ESV)** (ventricular volume at the end of the ejection period).

$$(SV) = (EDV) - (ESV)$$

Further, **ejection fraction** (EF) is defined as the stroke volume divided by the end-diastolic volume times 100%. Thus,

$$EF = (SV/EDV) \times 100\%$$

A normal value for ejection fraction in humans is typically greater than 55 percent.

To better understand the relationship between pressure and volume in the heart, we also introduce the concept of afterload. **Afterload** is the force against which the ventricle must contract to eject blood. This force is not really a single force, but can be thought of as the pressure that a chamber of the heart must generate in order to eject blood out of the chamber. Afterload is the tension produced in the heart fibers as the heart contracts. The afterload depends on blood pressure since pressure in the ventricle must exceed the blood pressure in the aorta to open the aortic valve. Afterload in the heart depends on the ventricular systolic pressure.

Starling's law of the heart relates the volume of the heart to the force of ejection that is available. **Starling's law** says that increasing the venous return to the heart stretches the ventricle, which in turn results in more forceful ejection of blood at the very next heartbeat. Within normal limits, longer myocardial fibers generate a larger force to expel blood from the ventricle.

Myocardial fiber length depends on the end-diastolic volume. End-diastolic volume, in turn, depends on the ventricular compliance and the end-diastolic pressure. Since cardiac output and stroke work are related to stroke volume, these quantities are all related to end-diastolic fiber length. In the same way, they are related to end-diastolic pressure, end-diastolic volume, and atrial pressure. Figure 2.23 shows how stroke work increases with increased end-diastolic fiber lengths. Simply said, if the heart doesn't fill well, it won't pump well.

Left ventricular (LV) stroke work can be calculated by integrating the product of pressure times volume at every pressure during the time of ejection. More simply, the stroke work can be estimated by multiplying the left ventricular stroke volume by the mean arterial pressure. That is,

LV stroke work = LV stroke volume × mean arterial pressure

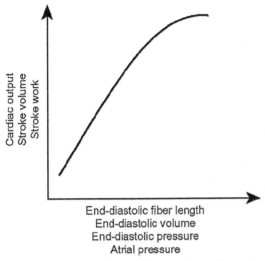

Figure 2.23 Increasing cardiac stroke work, stroke volume, and cardiac output as the end-diastolic fiber length changes, in accordance with Starling's law. (*From Tanner and Rhodes, 2003, p. 240.*)

Recall that over time, the left ventricular output must exactly match the right ventricular output. If more blood enters the atrium, a mechanism is necessary to increase the stroke volume and cause more blood to exit the ventricle.

It is useful here to establish the concept of **preload.** The preload establishes the fiber length before contraction. As with afterload, preload is not a single force. It is the tension produced in the heart fibers, in a chamber of the heart, because of filling.

Preload alone does not determine muscle fiber length. In heart disease, the same preload that would normally fill the ventricle and create a normal fiber length can result in a shorter fiber length and decrease the force of contraction, because of decreased volume compliance (increased stiffness).

2.8.2 Changes in contractility

Contractility can be thought of as the ability of a cardiac muscle fiber to contract at a given length. Other factors besides end-diastolic fiber length change the force of ventricular contraction. For example, when blood pressure decreases the autonomic nervous system releases the neurotransmitter norepinephrine. Norepinephrine increases the force of contraction for a given end-diastolic fiber length. Another determinant of stroke volume is afterload, the load against which the fibers must shorten. Afterload depends on arterial pressure.

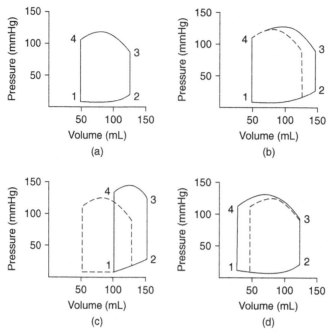

Figure 2.24 Pressure volume loops for (*a*) normal, (*b*) increased preload, (*c*) increased afterload, and (*d*) increased contractility. (*From Tanner and Rhodes, 2003, p. 243.*)

2.8.3 Ventricular performance

Figure 2.24 shows four different pressure-volume loops. Figure 2.24*a* is a normal pressure-volume loop. The points labeled 1, 2, 3, and 4 represent ventricular performance, with filling from 1 to 2, isovolumetric contraction from 2 to 3, ejection from 3 to 4, and ventricular relaxation from 4 to 1.

Note that the increased preload in Fig. 2.24*b* is a result of increased filling. In Fig. 2.24*c*, the increased aortic pressure results in an increased afterload. In that case, the ventricle does not empty as well. In Fig. 2.24*d*, the increased force of contraction causes more blood to be ejected, emptying the ventricle more completely and allowing more filling at the next stroke.

2.8.4 Clinical feature: congestive heart failure

In a patient with congestive heart failure, blood backs up on the atrial/venous side because the ventricle cannot keep up with the filling. There are many causes of congestive heart failure including myocardial infarction. Mitral valve failure would also lead to left congestive heart failure, and increased pulmonary venous pressure and pulmonary edema.

2.8.5 Pulsatility index

Pulsatility index (PI) is a unitless measure of the "amount of pulse" in a pressure or flow waveform. Pulsatility index is defined as the peak-to-peak value of the waveform divided by the mean value of the waveform. That is,

$$PI = \frac{\text{peak-to-peak value}}{\text{mean value}}$$

A large mean flow or pressure, with a relatively steady waveform, results in a small PI. A relatively small mean flow, combined with a relatively large peak-to-peak value, will result in a large PI. When reporting the PI, since it is unitless, it is important to report the type of waveform on which the PI is based, for example, pressure waveform PI or flow waveform PI.

2.8.6 Example problem: pulsatility index

The diagram shown in Fig. 2.25 is from a paper published in 2006 by Nagaoka and Yoshida. Find the PI for both the first-order and second-order arterioles. Do you expect the flow to become more or less pulsatile in the second-order arteriole compared to the first-order arteriole?

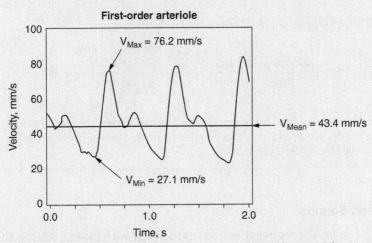

Figure 2.25 Velocity waveform from first- and second-order arterioles in a human retina. (*Modified from Nagaoka and Yoshida, 2006, p. 1114.*)

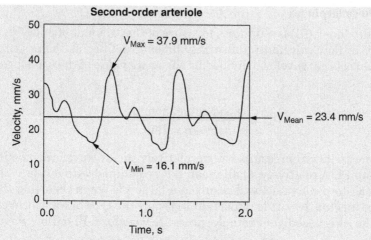

Figure 2.25 (*Continued*)

For the first-order arteriole, the maximum systolic-waveform velocity is 76.2 mm/s and the minimum diastolic velocity for the same pulse is 27.1 mm/s. The mean value for the given waveform is 43.4 mm/s.

We should state that we are calculating a velocity waveform PI, which will be equivalent to the flow waveform PI. The PI in the first-order arteriole can be calculated by

$$PI_1 = \frac{\text{peak-to-peak value}}{\text{mean value}} = \frac{76.2\,\frac{mm}{s} - 27.1\,\frac{mm}{s}}{43.4\,\frac{mm}{s}} = 1.13$$

The PI of the second-order arteriole can be calculated by

$$PI_2 = \frac{\text{peak-to-peak value}}{\text{mean value}} = \frac{37.9\,\frac{mm}{s} - 16.1\,\frac{mm}{s}}{23.4\,\frac{mm}{s}} = 0.93$$

As blood progresses from large arteries toward capillaries, the flow becomes less pulsatile. We would expect a smaller PI in the second-order arteriole than in the first-order arteriole.

2.9   Heart Sounds

Figure 2.19 also shows heart sounds associated with events during the cardiac cycle. Heart sound S1 is the "lub" associated with the closing of the AV valves. Heart sound S2 is the "dub" associated with the closing of the semilunar valves. Heart sounds are relatively low pitched. Most are between 10 and 500 Hz, and of a relatively low intensity. Some components of sounds S_1 and S_2 are produced by blood movement between atria and ventricles, by vibration of the ventricular walls, and even by vibration of the base of the aorta and pulmonary arteries. A third, less distinct heart sound may occur shortly after the opening of the AV valves. This sound can be heard most often in children or in adults with thin chest walls. A fourth sound may occur during atrial systole, when a small volume of blood is injected into the nearly filled ventricle. Once again, this less distinct sound is heard most often in children or in thin adults.

2.9.1 Clinical features

An increased end-systolic volume is typical in heart failure. In a patient with a failing heart, rapid early ventricular filling produces an audible third heart sound termed S_3, which occurs after S_2.

Late ventricular filling with atrial contraction is normally silent. In patients with hearts that have become stiffer, the end-diastolic blood flow is against a noncompliant ventricle, resulting in an audible fourth heart sound termed S_4. Heart sound S_4 would occur just prior to S_1.

2.9.2 Factors influencing flow and pressure

The relationship between blood flow and blood pressure is one of the main topics of this book. An engineer who is familiar with constant stroke-volume pumps might imagine that flow from the heart is a simple function of heart rate. An engineer who designs pressure controllers may think of flow to a given capillary bed as related directly to the hydraulic resistance in that bed. In fact, the heart can sometimes be modeled as a constant stroke-volume variable-speed pump, but the system containing the heart and circulatory system is a complex network with a control system that includes pressure inputs, flow inputs, as well as neural and chemical feedback.

The pulse pressure in the aorta can be thought of chiefly as a function of heart rate, peripheral resistance, and stroke volume.

Figure 2.26 shows the effect of heart rate on pressure. If peripheral resistance and stroke volume remain constant, an increase in heart rate causes the heart to pump more blood into the aorta over a fixed time period, and systolic blood pressure increases. At the same time, the increased rate of pumping leaves less time between heartbeats for the pressure in arteries to decrease, simultaneously increasing the diastolic pressure. As the heart rate increases with fixed peripheral resistance and stroke volume, cardiac output, or flow into the systemic circulation, increases.

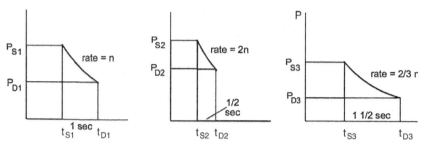

Figure 2.26 Aortic pressures as a function of heart rate. P_S is the systolic pressure, and P_D is the diastolic pressure. Picture 2, the middle picture, shows the fastest heart rate, and the picture on the right, picture 3, shows the slowest heart rate. (*Modified from Selkurt, 1982, Figure 8.26.*)

On the other hand, if heart rate and peripheral resistance are constant, but stroke volume changes, pressure and flow are also affected. Figure 2.27 shows the effect of stroke volume on pressure. An increase in stroke volume increases pressure because of the extra volume of blood that is pumped into the aorta during each heartbeat. However, a healthy human would not normally experience a case of constant heart rate and constant peripheral resistance with increasing stroke volume. As shown in the figure, under resting conditions, heart rate drops and peripheral resistance increases. The result is that a large stroke volume, as might be seen in well-trained athletes, increases systolic pressure or perhaps maintains it at a steady level, while the decreased heart rate allows more time between beats and the diastolic pressure decreases. A very small stroke volume, as might be seen in a patient with severe heart damage, would normally result in an increased heart rate as the cardiovascular system attempts to maintain cardiac output, although a very small amount of blood is ejected from the left ventricle during each heartbeat.

The third parameter that controls blood pressure and flow is peripheral resistance. This peripheral resistance comes chiefly from the resistance found in capillary beds, and the more capillaries that are open, the lower the resistance will be. If heart rate and stroke volume remain constant,

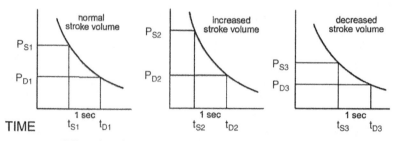

Figure 2.27 Effect of stroke volume on pressure. P_S represents pressure at systole, and P_D represents pressure at diastole. (*Modified from Selkurt, 1982, Figure 8.26.*)

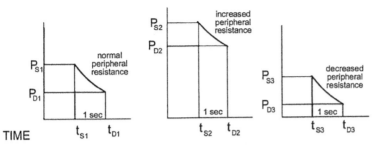

Figure 2.28 Aortic pressure as a function of peripheral resistance while heart rate and stroke volume are held constant. (*Modified from Selkurt, 1982, Figure 8.26.*)

as the need for oxygenated blood increases in peripheral tissues, more capillaries open to allow more blood to flow into the tissue; this corresponds to a decrease in peripheral resistance. Figure 2.28 shows that blood pressure increases with increasing peripheral resistance when heart rate and stroke volume are held constant.

Heart rate, stroke volume, and peripheral resistance are the primary factors that control the relationship between blood pressure and flow, or cardiac output, but there are other factors which contribute. For example, in the case of diseased heart valves, blood leakage backward through the valve into the heart can alter the relationship. This type of leakage through a valve is known as regurgitance. Regurgitance in heart valves has a similar effect to that of decrease in stroke volume. Blood which has been ejected into the aorta can flow backward into the left ventricle through a faulty aortic valve, resulting in a decreased effective stroke volume.

Arterial compliance, a characteristic which is related to vessel stiffness, can also affect the pressure flow relationship. Very stiff arteries increase the overall hydrodynamic resistance of the system. This increase in resistance causes a decrease in blood flow to the peripheral tissue. The complex control system in the cardiovascular system will result in a compensation that increases heart rate, and therefore increases blood pressure to compensate for the increased resistance.

2.10   Coronary Circulation

The blood supply to the heart is known as the **coronary circulation.** Because heart muscle requires a blood supply in order to be able to provide blood to the rest of the body, this system, which is relatively small by blood volume, provides a very important function. Approximately, one-third of people in the countries of the western world die as a result of coronary artery disease. Almost all elderly people have some impairment of the coronary circulation.

The resting coronary blood flow in humans is around 225 mL/min, which is a bit less than 5 percent of cardiac output. During strenuous exercise, cardiac output in young healthy people increases four- to sevenfold. Because the heart also provides this increased flow at a higher arterial pressure, the work of the heart is increased six to nine times over the resting work output. The coronary blood flow increases only six- to ninefold to supply oxygen and nutrients needed by the cardiac muscle, during this increased workload.

The main coronary artery arises from the root of the aorta. It then branches into the left and right coronary arteries. Seventy-five percent of the coronary blood supply goes through the left coronary artery and 25 percent through the right coronary artery.

The left coronary artery supplies blood to the left lateral portion of the left ventricle and the anterior (front) portion of the ventricular septum. The ventricular septum is the wall between the left and right ventricles. The **left coronary artery** then branches into the **anterior descending coronary artery** and the **circumflex coronary artery.** Those two branches provide blood supply to the left ventricle.

The **right coronary artery** supplies the right ventricle and the posterior (back) portion of the ventricular septum. The right coronary artery also supplies blood to the specialized muscle fibers of the sinoatrial (SA) node, which is located at the top-rear-right of the right atrium and is the primary electrical pacemaking site of the heart.

Blood flows through the capillaries of the heart and returns through the cardiac veins to the coronary sinus. The coronary sinus is the main drainage channel of venous blood from the heart muscle. It has the appearance of a groove on the back surface of the heart, between the left atrium and ventricle, and drains into the right atrium. Most of the coronary blood flow from the left ventricle returns to the coronary circulation through the coronary sinus. Most coronary blood flow to the right ventricle returns through the small anterior cardiac veins that flow directly into the right atrium.

2.10.1   Control of the coronary circulation

Control of the coronary circulation is almost entirely local. The coronary arteries and arterioles vasodilate (increase in diameter) in response to the cardiac muscle's need for oxygen. Whenever muscular activity of the heart increases, blood flow to the heart increases. Decreased activity also causes a decrease in coronary blood flow. After blood flows through the coronary arteries, about 70 percent of the oxygen in the arterial blood is removed as it passes through the heart muscle. This means that there is not much oxygen left in the coronary venous blood that could have been removed in more strenuous activity. Therefore, it is necessary that coronary flow increase as heart work increases.

2.10.2 Clinical features

When blockage of a coronary artery occurs gradually over months or years, collateral circulation can develop. This collateral circulation allows adequate blood supply to the heart muscle to develop through an alternative path. On the other hand, since atherosclerosis is typically a diffuse process, blockage of other vessels will also eventually occur. The end result is a devastating and sometimes lethal loss of heart function because the blood flow was dependent on a single collateral branch.

2.11 Microcirculation

Microcirculation is the flow of blood in the system of smaller vessels of the body whose diameters are 100 μm or less. Virtually all cells in the body are spatially close to a microvessel. There are tens of thousands of microvessels per gram of tissue in the human body. These include **arterioles, metarterioles, capillaries,** and **venules.** The microcirculation is involved chiefly in the exchange of water, gases, hormones, nutrients, and metabolic waste products between blood and cells. A second major function of the microcirculation is regulation of vascular resistance.

Normally, all microvessels other than capillaries are partially constricted by contraction of smooth muscle cells. If that is so, then why are the capillaries not constricted in diameter? It is because capillaries do not have smooth muscle cells in the wall. Precapillary sphincters can close off some capillaries, but capillaries cannot change their diameters.

If all of your smooth muscle cells relaxed instantaneously, you would faint immediately. Such relaxation would allow a large increase in volume of blood in the cardiovascular system, dropping blood pressure precipitously. In skeletal muscle and other tissues, a large number of capillaries can remain closed for long periods due to the contraction of the **precapillary sphincter.** This mechanism is a part of the regulation of vascular resistance.

These capillaries provide a reserve flow capacity and can open quickly in response to local conditions, such as a fall in the partial pressure of oxygen, when additional flow is required.

2.11.1 Capillary structure

Capillaries are thin tubular structures whose walls are one cell layer thick. They consist of highly permeable endothelial cells. Figure 2.29 shows the various layers of a capillary. In the peripheral circulation, there are about 10 billion capillaries with an average length of about 1 mm. The estimated surface area of all the capillaries in the entire system is about 500 m^2, and the total volume of blood contained in those

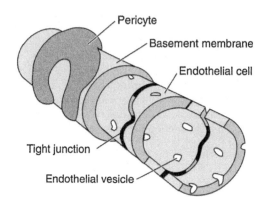

Pericyte

Basement membrane

Endothelial cell

Tight junction

Endothelial vesicle

Figure 2.29 Structure of a human capillary. (*From Tanner and Rhodes, 2003, p. 264.*)

capillaries is about 500 mL. Capillaries are so plentiful that it is rare that any cell in the body is more than 20 μm away from a capillary.

Consider the structure of blood vessels in terms of branching "generations." If the aorta branches or bifurcates into two branches, those branches may be considered the second generation. Arteries branch six or eight times before they become small enough to be **arterioles.** Arterioles then branch another two to five times, for a total of eight to thirteen generations of branches within arteries and arterioles. Finally, the arterioles branch into metarterioles and capillaries that are 5 to 9 μm in diameter.

In the last arteriole generation, blood flows from the arterioles into the metarterioles, which are sometimes called terminal arterioles. From there, the blood flows into the capillaries.

There are two types of capillaries. The first type is the relatively larger **preferential channel;** the second type is the relatively smaller **true capillary.** At the location where each true capillary begins from a metarteriole, there is smooth muscle fiber, known as the precapillary sphincter, surrounding the capillary. The precapillary sphincter can open and close the entrance to a true capillary, completely shutting off blood flow through this capillary.

After leaving the capillaries, most blood returns to **venules** and then eventually back into the veins. However, about 10 percent of the fluid leaving the capillaries enters the lymphatic capillaries and returns to the blood through the lymphatic system.

Arteriolar wall structure.   Arterioles are tubes of endothelial cells surrounded by a basement membrane, a single or double layer of smooth muscle cells, and a thin outer layer of connective tissue. In most organs, arteriolar smooth muscle cells operate at about half their maximum length.

2.11.2 Capillary wall structure

A capillary is a tube of endothelial cells surrounded by a basement membrane. The wall of a capillary is one cell layer thick and consists of endothelial cells surrounded on the outside by a basement membrane. Pericytes (Rouget cells) are connective tissue and may be a primitive form of vascular smooth muscle cells, and add structural integrity to the capillary. Precapillary sphincters are small helical twists of smooth muscle around the smallest arterioles (metarterioles). These sphincters are very responsive to the local or intrinsic effects like the concentration of oxygen or carbon dioxide.

The wall thickness of a capillary is about 0.5 μm. The inside diameter is approximately 4 to 9 μm which is barely large enough for an erythrocyte, or red blood cell, to squeeze through. In fact, erythrocytes, which are approximately 8 μm in diameter, must often fold in order to squeeze through the capillary.

Capillary walls have intercellular clefts, which are thin slits between adjacent endothelial cells. The width of these slits is only about 6 to 7 nm (6 to 7 $\times$ 10^{-9} m) on the average. Water molecules and most water-soluble ions diffuse easily through these pores.

The pores in capillaries are of different sizes for different organs. For example, intercellular clefts in the brain are known as "tight junctions." These pores allow only extremely small molecules like water, oxygen, and carbon dioxide to pass through the capillary wall. In the liver, the opposite is true. The intercellular clefts in the liver are wide open and almost anything that is dissolved in the plasma also goes through these pores.

Because a single capillary has such a small diameter, each capillary taken alone has a very high hydraulic resistance. However, when you add together many capillaries in parallel, the capillaries account for only about 15 percent of the total resistance in each organ.

2.11.3 Pressure control in the microvasculature

Arterioles regulate vascular pressures and microvascular resistance. A constant "conflict" exists between preserving arterial pressure by increasing arteriolar resistance and allowing all regions to receive sufficient perfusion to provide oxygen to the tissue. By decreasing resistance in small vessels, more blood flows and oxygen perfusion increases. By decreasing resistance in arterioles and capillaries, however, system blood pressure falls.

The normal pressure in an arteriole is typically 30 to 70 mmHg, while the pressure in a venule is 10 to 16 mmHg. Large arteries are simply conduits to allow blood transfer between locations. These large vessels have very low resistance and do not play a significant role in pressure regulation. Small arteries (0.5 to 1 mm in diameter) control 30 to 40 percent of total vascular resistance, and arterioles (<500 μm in diameter) combined with those small arteries make up 70 to 80 percent of total vascular resistance. Nearly all of the rest of the resistance (20 to 30 percent) comes from capillaries and venules.

Constriction of smooth muscle cells maintains the resistance in these resistance vessels. Release of norepinephrine from the sympathetic nervous system contributes to smooth muscle constriction. When fully relaxed, the diameter of the vessel surrounded by smooth muscle can nearly double. If the vessel diameter doubles, the resistance over a fixed length of vessel increases by $2^4 = 16$ times.

Arterioles possess significant smooth muscle cells and can dilate 60 to 100 percent from their resting diameter. Arterioles can also constrict 40 to 50 percent from their resting diameter for long periods.

Only approximately 10 percent of muscle capillaries are open at rest. Therefore, muscle is capable of increasing its rate of bulk flow tenfold as the need for oxygen increases. Since humans are not capable of increasing their cardiac output tenfold (the human heart can achieve a maximal five-fold increase), shutting down other capillary beds is a necessary pressure control mechanism. That is, you can increase the blood flow to some muscle tenfold so long as not all muscles in the body try to do that simultaneously.

2.11.4 Diffusion in capillaries

Lipid-soluble molecules like O_2 and CO_2 easily pass through the lipid components of **endothelial cell membranes.** Water-soluble molecules like glucose must diffuse through water-filled pathways in the capillary wall between adjacent endothelial cells. The **water-filled pathways** are known as pores. In humans, adjacent endothelial cells are held together by **tight junctions**, which have occasional gaps.

Capillaries of the brain and spinal cord allow only the smallest water-soluble molecules to pass through. Pores are partially filled with sub-micron fibers (fiber matrix) that act as a filter.

Most pores allow passage of 3- to 6-nm molecules. This pore size allows the passage of water, inorganic ions, glucose, and amino acids. Pores exclude passage of large molecules like **serum albumin** and **globular proteins.** A limited number of large pores exist. It is possible that these larger pores are actually defects in the cell wall. Large pores allow passage of virtually all large molecules. Because of these large pores and their associated leakage, nearly all serum albumin molecules leak out of the cardiovascular system each day.

2.11.5 Venules

Venules are endothelial tubes that are usually surrounded by a mono-layer of vascular smooth muscle cells. Smooth muscle cells in venules are much smaller in diameter, but longer than those of arterioles.

The smallest venules are more permeable than capillaries. A significant amount of the diffusion that occurs in the microcirculation, therefore, occurs in the venules.

Tight junctions are more frequent and have bigger pores. Therefore, it is probable that much of the exchange of large water-soluble molecules occurs as blood passes through small venules.

The venous system acts like a huge blood reservoir, and at rest, approximately two-thirds of blood is located in the venous system. By changing venule diameter, volume of blood in tissue can change up to 20 mL/kg. For a 70-kg (154 lb) person, the volume of blood ready for circulation can change by nearly 1.5 L.

2.12 Lymphatic Circulation

Lymphatic vessels are endothelial tubes within the tissues. They carry fluid, serum proteins, lipids, and foreign substances from the interstitial spaces back to the circulatory system.

Lymphatic vessels begin as blind-ended lymphatic bulbs. The wall of a lymph capillary is constructed of endothelial cells that overlap one another. The pressure in the interstitial fluid surrounding the capillary pushes open the overlapping cells of the lymphatic bulbs. The movement of fluid from the surrounding tissue into the lymphatic vessel lumen is passive. That is, there are no active pumps in the lymphatic system. Once the interstitial fluid enters a lymph capillary, it is referred to as lymph.

A compression/relaxation cycle causes the lymphatic vessels to "suck up" lymph. Lymph valves in lymphatic vessels make this possible. A volume of fluid equal to the total plasma volume in a single human is filtered from the blood to the tissues each day. It is critical that this fluid is returned to the venous system by lymph flow each day. The lymphatic vasculature collects approximately 3 L/day of excess interstitial fluid and returns it to the blood. The lymph system is responsible for returning plasma proteins to the blood.

Lymphatic pressures are only a few mmHg in the bulbs and the smallest lymphatic vessels, but can be as high as 10 to 20 mmHg during the contraction of larger lymphatic vessels. Lymph valves make possible this apparent progression or flow from low pressure toward high pressure.

Review Problems

1. According to Poiseuille's law, flow in a large artery is proportional to

    (a) radius × viscosity
    (b) radius^4 × (pressure gradient)/viscosity
    (c) radius^4/(pressure gradient)
    (d) radius^2/viscosity
    (e) viscosity/radius

2. According to the Reynolds number, turbulence is more likely with
   (a) increasing viscosity
   (b) decreasing tube diameter
   (c) decreasing velocity
   (d) increasing velocity
   (e) increasing pressure

3. Decreased aortic compliance means
   (a) decreased stiffness, resulting in increased resistance
   (b) increased change in cross-sectional area for a given pressure increase
   (c) increased stiffness, resulting in increased mean aortic pressure for the same cardiac output
   (d) increased stiffness, resulting in a decrease in mean aortic pressure for same cardiac output
   (e) decreased stiffness, resulting in a decrease in mean aortic pressure

4. Cardiac output is approximately proportional to
   (a) aortic pressure divided by systemic vascular resistance
   (b) systemic vascular resistance times aortic pressure
   (c) systemic vascular resistance times right atrial pressure
   (d) right ventricular pressure divided by systemic vascular resistance
   (e) aortic pressure times stroke volume

5. If the heart were stopped, after due time, the pressure in the entire system circulation would come to equilibrium at some pressure greater than 0. This equilibrium pressure is known as
   (a) mean aortic pressure
   (b) diastolic pressure
   (c) mean circulatory filling pressure
   (d) vena cava pressure
   (e) central venous pressure

6. For the following ECG, find the mean electrical axis.

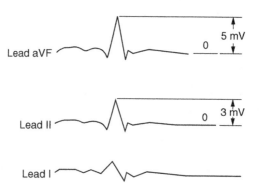

7. The QRS complex of the ECG is associated with

(a) atrial depolarization
(b) atrial repolarization
(c) bundle branch block
(d) ventricular repolarization
(e) ventricular depolarization

8. Find the mean electrical axis for the following ECG:

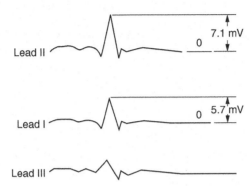

9. Excitation of the heart normally begins

(a) at the sinoatrial node
(b) at the atrioventricular node
(c) in the bundle of His
(e) in the left ventricular myocardium
(f) in the Purkinje system

10. The following ECG shows a patient with

(a) atrial flutter
(b) atrial tachycardia
(c) ventricular tachycardia
(d) right bundle branch block
(e) left bundle branch block

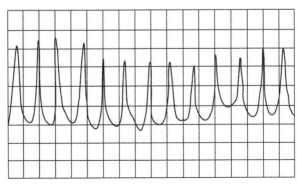

Modified from Selkurt, 1982, p. 287, Figure 8.13G.

11. When are atrioventricular valves open?

   (a) Atrioventricular valves are open only during atrial systole.
   (b) Atrioventricular valves are open during ventricular systole.
   (c) Atrioventricular valves are open during atrial systole as well as during ventricular diastole.
   (d) Atrioventricular valves are open during atrial systole and the first half of ventricular systole.
   (e) Atrioventricular valves are open when the flow rate through the vena cava is greater than the flow rate through the aorta.

12. A patient has an end-diastolic left ventricular volume of 100 mL and a left ventricular stroke volume of 50 mL. The ejection fraction is

   (a) 20%
   (b) 50%
   (c) 33%
   (d) 75%
   (e) undeterminable due to lack of information

13. The same patient in problem 12 has a heart rate of 60 bpm. The cardiac output in this patient with a 100-mL end-diastolic left ventricular volume and 50-mL left ventricular stroke is

   (a) 3 L/min
   (b) 4 L/min
   (c) 5 L/min
   (d) 6 L/min
   (e) undeterminable due to lack of information

14. The figure below shows a normal pressure-volume loop for a patient. Which of the three figures (a), (b), or (c) indicate an increased preload?

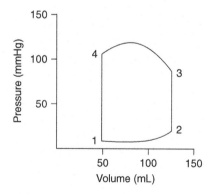

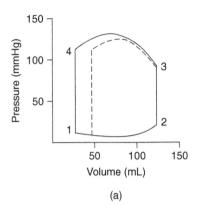

(a)

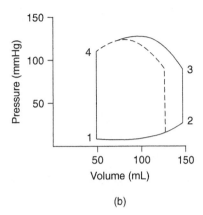

(b)

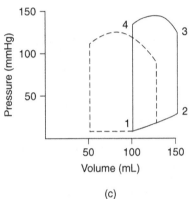

(c)

(d) none of the pictures shows increased preload

15. In the pressure-volume loop shown at the left, point 4 represents

(a) closing of the aortic valve
(b) opening of the mitral valve
(c) closing of the mitral valve
(d) opening of the aortic valve
(e) beginning of ventricular systole

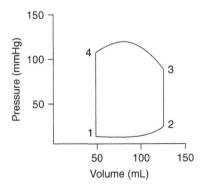

16. How much of the systemic blood volume is located on the venous side of the circulation?

(a) 25%
(b) 40%
(c) 50%
(d) 60%
(e) 75%

17. Central blood volume affects cardiac output in the following way:

(a) An increase in central blood volume decreases preload and cardiac output.
(b) An increase in central blood volume increases preload and decreases cardiac output.
(c) A decrease in central blood volume decreases preload and cardiac output.
(d) A decrease in central blood volume increases preload and cardiac output.
(e) Cardiac output is not affected by a change in central blood volume.

18. The relationship between arterial compliance and arterial pressure is best described by

(a) Arterial compliance increases as arterial pressure increases.
(b) Arterial compliance decreases as arterial pressure increases.
(c) Arterial pressure and arterial compliance are unrelated.
(d) Arterial compliance multiplied by arterial pressure equals heart work.
(e) Low arterial compliance can result in lowered arterial pressure.

19. Systemic vascular resistance is most influenced by

(a) radius of the aorta
(b) radius of medium-sized arteries
(c) blood viscosity
(d) percentage of "open" capillaries
(e) radius of arterioles

20. In capillary exchange, lipid-soluble molecules, like oxygen and carbon dioxide, pass from the blood to the tissues through

(a) water-filled pathways called pores
(b) tight junctions
(c) the lipid components of endothelial cell membranes
(d) endothelial vesicles
(e) large pores

21. Regulation of vascular resistance in the microcirculation occurs primarily in

(a) small arteries
(b) arterioles
(c) capillaries
(d) venules
(e) small veins

22. Lymphatic pressures are only a few mmHg in the bulbs and smaller lymphatic vessels. The lymphatic pressure during contraction of large lymph vessels is

(a) −5 to −1 mmHg
(b) −1 to 0 mmHg
(c) 0 to 1 mmHg
(d) 10 to 20 mmHg
(e) 50 to 75 mmHg

23. Blood flows through what percent of skeletal muscle capillaries at rest?

(a) 10%
(b) 25%
(c) 50%
(d) 75%
(e) 90%

24. Capillary pores are the water-filled pathways that allow passage of water-soluble molecules from the blood to the tissue. Pores typically exclude molecules like

(a) inorganic ions
(b) glucose
(c) amino acids
(d) albumin and globular proteins
(e) all of the above

Bibliography

Herse B, Baryalei M, Wiegand V, and Autschback R. Fractured strut of Björk-Shiley 70 degrees convexo-concave mitral valve prosthesis found in left coronary artery. *Thorac Cardiovas Surg,* 1993; 41:77–79.

Guyton AC and Hall JE. *Textbook of Medical Physiology,* 10th ed. Saunders, Philadelphia; 2000.

Lingappa VR and Farey K. *Physiological Medicine.* McGraw-Hill, New York; 2000.

Milnor WR. *Cardiovascular Physiology.* Oxford University Press, New York; 1990.

Nagaoka T and Yoshida A. Noninvasive evaluation of wall shear stress on retinal microcirculation in humans, *Invest Ophthalmol Vis Sci,* 2006; 47(3):1114.

Selkurt EE. *Basic Physiology for the Health Sciences,* 2nd ed. Little, Brown and Company, Boston; 1982.

Tanner GA and Rhodes RA. *Medical Physiology,* 2nd ed. Little, Brown and Company, Boston; 2003.

Webster JG (ed.). *Medical Instrumentation,* 3rd ed. Houghton Mifflin, Boston, MA; 1998.

Yoganathan AP. Cardiac valve prostheses, in Bronzino JD (ed.), *The Biomedical Engineering Handbook.* CRC Press, Boca Raton, FL; 1995:pp. 1847–1870.

Pulmonary Anatomy, Pulmonary Physiology, and Respiration

3.1 Introduction

The three most important functions of the respiratory system are gas exchange, acid-base balance, and the production of sound. The most basic activity of the respiratory system is to supply oxygen for metabolic needs and to remove carbon dioxide. Carbon dioxide, which is carried in the hemoglobin of our red blood cells, is exchanged in the lungs for oxygen, which is also carried in the hemoglobin. Fresh air, which we continuously inspire into the lungs, is exchanged for air that has been enriched with carbon dioxide. Increases in carbon dioxide lead to increases in hydrogen ion concentration resulting in lower pH or more acidity. The respiratory system aids in acid-base balance by removing CO_2 from the body. Phonation, or the production of sounds by air movement through vocal cords, is the third important function of the respiratory system. When you speak, the muscles of respiration cause air to move over the vocal cords and through the mouth resulting in intelligible sounds.

To understand the path of air through the respiratory system, consider the system components, which include two lungs, a trachea, bronchi, bronchioles, alveoli, and their associated blood vessels. These components are shown in Fig. 3.1. Air enters the respiratory system through the mouth or nose. After being filtered and heated, the air passes through the pharynx and the larynx. It then enters the trachea from where it passes to the bronchial tree. There are three generations of bronchi, which are conducting airways greater than about 0.5 cm in diameter. Bronchioles are also conducting airways that have diameters greater than approximately 0.05 cm. There are approximately 13 generations of bronchioles, with the last generation being known as terminal bronchioles.

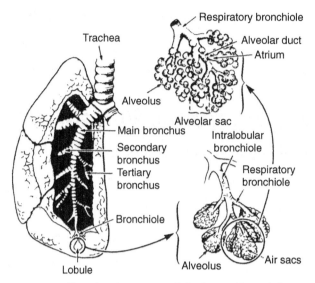

Figure 3.1 Functional anatomy of the lung. (*From Selkurt, 1982, Figure 9.2, p. 327.*)

Beyond the terminal bronchioles, air enters the transitional zone. There are several generations of respiratory bronchioles, just beyond the bronchi. Respiratory bronchioles have a diameter of approximately 0.05 cm and a length of about 0.1 cm. At the ends of the respiratory bronchioles, air passes into alveolar ducts and finally into alveolar sacs. Alveolar ducts and alveolar sacs make up the respiratory zone of the respiratory system. Figure 3.1 shows the functional anatomy of the lungs. At the left, you can see the gross structure of the right lung and the conducting airways. At the upper right, the branching of a bronchiole can be seen. Finally at the lower right, at highest magnification, respiratory bronchioles, alveolar ducts, alveolar sacs, and alveoli are shown.

3.1.1   Clinical features: hyperventilation

Oxygen and carbon dioxide in blood are carried in hemoglobin, a substance inside red blood cells. Hemoglobin in oxygenated blood that is returning from areas of the lung where gas exchange has occurred is normally saturated. Therefore, hyperventilation of room air will not add much oxygen to the bloodstream. If a patient suffers from low oxygen due to a mismatch in ventilation and perfusion, the patient must breathe air with a higher percentage of oxygen than that which occurs in room air.

(Lingappa, pp 381)

3.2   Alveolar Ventilation

Alveolar ventilation is the exchange of gas between the alveoli and the external environment. It can be measured as the volume of fresh air entering (and leaving) the alveoli each minute. Oxygen from the atmosphere enters the lungs through this ventilation and carbon dioxide from the venous blood returns to the atmosphere. Physicians and biomedical engineers often discuss alveolar ventilation in terms of standard lung volumes. These standard lung volumes are represented graphically in Fig. 3.2.

3.2.1   Tidal volume

Tidal volume (TV) is the volume of ambient air entering (and leaving) the mouth and nose per minute during normal, unforced breathing. Normal TV for a healthy 70-kg adult is approximately 500 mL per breath, but this value can vary considerably during exercise.

3.2.2   Residual volume

Residual volume (RV) is the volume of air left in the lungs after maximal forced expiration. This volume is determined by a balance between muscle forces and the elastic recoil of the lungs. RV is approximately 1.5 L in a healthy 70-kg individual. RV is much greater for individuals with emphysema.

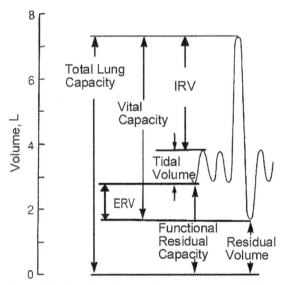

Figure 3.2   Standard lung volumes as measured by a spirometer.

3.2.3 Expiratory reserve volume

Expiratory reserve volume (ERV) is the volume of air expelled from the lungs during a maximal forced expiration that begins at the end of normal tidal expiration. The ERV plus the RV combine to make up the functional residual capacity. Normal ERV for our standard 70-kg individual is approximately 1.5 L.

3.2.4 Inspiratory reserve volume

Inspiratory reserve volume (IRV) is the volume of air that is inhaled during forced maximal inspiration beginning at the end of normal tidal inspiration. This volume depends on muscle forces and also on the elastic recoil of the chest wall and the elastic recoil of the lungs. The IRV of a healthy 70-kg individual is approximately 2.5 L.

3.2.5 Functional residual capacity

Functional residual capacity (FRC) is the volume in the lungs at the end of normal tidal expiration. FRC depends on the equilibrium point at which the elastic inward recoil of the lungs balances the elastic outward recoil of the chest wall. FRC consists of the sum of the RV and the ERV and can be represented by Eq. (3.1). The FRC of a healthy 70-kg adult is approximately 3 L.

$$FRC = RV + ERV \tag{3.1}$$

3.2.6 Inspiratory capacity

Inspiratory capacity (IC) is the volume of air taken into the lungs during a maximal inspiratory effort that starts at the end of a normal TV expiration. IC is the sum of TV and IRV and can be represented by Eq. (3.2). The IC of our healthy 70-kg adult is about 3 L.

$$IC = IRV + TV \tag{3.2}$$

3.2.7 Total lung capacity

Total lung capacity (TLC) is the volume of air in the lungs after a maximal inspiratory effort. The strength of contraction of inspiratory muscles and the inward elastic recoil of the lungs and the chest wall determine the TLC. The TLC is also equal to the sum of the IRV, the TV, the ERV, and the RV. A typical value for TLC of a healthy

70-kg male is about 6 L. An equation representing the TLC may be written by:

$$TLC = IRV + TV + ERV + RV \qquad (3.3)$$

3.2.8   Vital capacity

Vital Capacity (VC) is the volume of air expelled from the lung during a maximal expiratory effort after a maximum forced inspiration. The VC is the difference between the TLC and the RV. VC in a normal healthy 70-kg adult is about 4.5 L. Equation 3.4 shows an equation for VC.

$$VC = TV + IRV + ERV \qquad (3.4)$$

3.3   Ventilation-Perfusion Relationships

Ventilation is the act of supplying air into the lungs and **perfusion** is the pumping of blood into the lungs. In this book we will use $\dot{V}$ to represent the airflow rate associated with ventilation and $\dot{Q}$ to represent the blood flow rate associated with perfusion. The **ratio of ventilation to perfusion** is important for lung function and is represented as the ventilation/perfusion ratio as shown in Eq. (3.5).

$$\text{ventilation-perfusion ratio} = \frac{\dot{V}}{\dot{Q}} \qquad (3.5)$$

Pulmonary artery smooth muscle constricts the vessels of the pulmonary capillary beds in response to **hypoxia** or low oxygen. This type of vasoconstriction is one of the most important parameters that determine pulmonary blood flow. In other capillary beds within the body, smooth muscle vasodilates in response to tissue hypoxia, improving perfusion.

Resting ventilation is about 4 to 6 L/min. Resting pulmonary artery blood flow is about 5 L/min. At rest, therefore, the ventilation/perfusion ratio is about 0.8 to 1.2. Figure 3.3 shows a schematic representing a ventilation/perfusion ratio of 1 with a ventilation rate and perfusion rate both equal to 5 L/min.

3.4   Mechanics of Breathing

For the normal physiological case of breathing, air flows into the lungs when the alveolar pressure drops below the pressure of the surrounding ambient air. This is known as **"negative pressure breathing"**

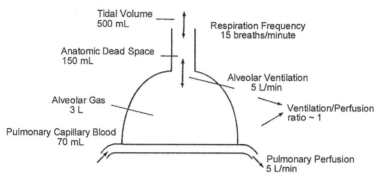

Figure 3.3 Pulmonary volumes and flows showing a ventilation/perfusion ratio of 1. (*From West, 1990 Figure 2.1, p. 12.*)

because the pressure in the alveoli must be negative with respect to the surrounding air. This negative pressure in the alveoli is caused by muscle contractions that increase the volume of the lung causing the alveoli to expand.

Transmural pressure gradient is defined by the difference in pressure between atmospheric air and the air in the alveoli. As the transmural pressure gradient increases, the alveoli expand.

Intrapleural pressure, which is also known as intrathoracic pressure, is caused by the mechanical interaction between the lung and chest wall. When all muscles of respiration are relaxed, left to themselves, the lungs have a tendency to collapse whereas the chest wall tends to expand. This causes the intrapleural pressure to drop, and this resulting negative pressure has the effect of holding the lung and the chest wall in close contact.

The primary muscles of breathing are the diaphragm, the external intercostals, and the accessory muscles. The diaphragm is a dome-shaped sheet of muscle with an area of about 250 cm^2. The diaphragm separates the abdominal cavity from the thoracic cavity.

During **eupnea** or normal quiet breathing, in the supine position the diaphragm is responsible for two-thirds of the air entering the lungs.

3.4.1 Muscles of inspiration

The rib muscles (external intercostals) raise and enlarge the rib cage when contracted. The diaphragm and the rib muscles contract simultaneously during inspiration. If they did not, the contraction of the external intercostals could cause the diaphragm to be pulled upward.

Accessory muscles are not used in normal quiet breathing. They are used, however, during heavy breathing, as in exercise, and during the inspiratory phase of sneezing and coughing. An example of an accessory muscle is the sternocleidomastoid, which elevates the sternum to increase the anteroposterior (front to back) and the transverse (side to side) dimensions of the chest.

3.4.2 Muscles of expiration

Expiration is passive during quiet breathing and no muscle contraction is required. The elastic contraction, or recoil, of the alveoli due to alveolar stiffness is enough to raise the alveolar pressure above atmospheric pressure, the condition required for expiration.

Active expiration occurs during exercise, speech, singing, and the expiratory phases of coughing and sneezing. Active expiration may also be required due to pathologies such as emphysema. Muscles of active expiration are the muscles of the abdominal wall including the rectus abdominis, external and internal obliques, transverse abdominis, and internal intercostals.

3.4.3 Compliance of the lung and chest wall

The slope of the pressure-volume (P-V) curve of the lung is known as **lung compliance**. This volume compliance can be written as dV/dP, representing a change in volume per change in pressure. Compliance has the units of m^5/N (m^3/Pa) and is inversely related to elasticity, or lung stiffness.

The **P-V curve** for the lung is different for inspiration than it is for expiration. The difference in volume for a given pressure upon inspiration versus the same pressure upon expiration is known as hysteresis. At low lung volumes, the lung is stiffer (has a lower compliance) during inspiration than during expiration. At high lung volumes, the lung is less stiff (has a higher compliance) during inspiration.

3.4.4 Elasticity, elastance, and elastic recoil

Compliance is a measure of lung distensibility, change in length per change in tension or change in volume per change in pressure. Elastic recoil, on the other hand, is the ability of the inflated lungs to return to their normal resting volume. Compliance is inversely related to elastic recoil.

Stiffness is the ratio of force to a measure of the deformation that the force produces. For example, a spring which stretches 1 mm in response to a force of 100 N would be said to have a stiffness of 100 N/mm. In the same way, a vessel which expands by 100 mL in response to a pressure increase of 20 kPa could be said to have a stiffness of 0.2 kPa/mL. Compliance is merely the reciprocal of stiffness, so the spring would have a compliance of 0.01 mm/N and the vessel's compliance would be 5 mL/kPa.

High stiffness means high resistance to stretch or inflation. Therefore, elastic recoil is directly proportional to lung stiffness. When the elastic recoil of the alveoli is compromised, as in the case of emphysema, then it becomes difficult for the person to forcibly exhale.

Elastance is a medical term which is used to describe the stiffness of hollow organs. It is the ability to resist deformation or the ability to return to its original shape after deformation. The concept of elastance and elasticity can be confusing to students sometimes. Using this concept, a steel rod has more elastance or is more elastic than a rubber band. Again, this only means that the steel rod is stiffer, while the rubber band is more compliant.

3.4.5   Example problem: compliance

From the P-V curve shown, calculate the compliance of the lung during inspiration between points A and B in m^5/N.

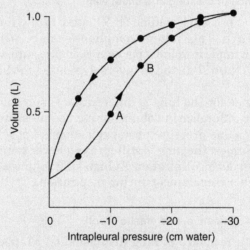

(Modified from West, 1990, p. 90, Figure 7.3.)

Solution   A straight line through points A and B is extended on the plot below. The slope of the line is

$$C = \frac{\Delta V}{\Delta P} = \frac{1\ \text{L}}{20\ \text{cm H}_2\text{O}} \ \frac{1000\ \text{cm}^3}{1\ \text{L}} \left(\frac{1\ \text{m}}{100\ \text{cm}}\right)^3 \left(\frac{1.35\ \text{cm H}_2\text{O}}{1\ \text{mmHg}}\right) \left(\frac{1\ \text{mmHg}}{133.3\ \text{N/m}^2}\right)$$

$$C = 50\ \frac{\text{mL}}{\text{cm H}_2\text{O}} = 5.1 \times 10^{-3}\ \frac{\text{m}^5}{\text{N}}$$

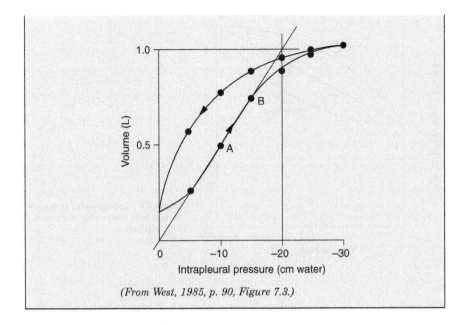

(From West, 1985, p. 90, Figure 7.3.)

3.5 Work of Breathing

The rate and depth at which one breathes under normal circumstances, is managed to minimize the amount of work that is done. If you try to breathe rapidly and shallowly for an extended period of time, you can transfer the necessary oxygen, but will rapidly grow tired from the effort. Work for a system that executes a cyclic process with only expansion and compression can be modeled by Eq. (3.6) (Wark and Richards, 1999).

$$W_{a \to b} = \int_a^b P \, dV \qquad (3.6)$$

In Eq. (3.6), W represents the work done between points a and b. P represents the pressure inside the system (in this case the lung). The volume of the system is represented by V. Figure 3.4 shows a typical P-V curve during breathing. The work term can be thought of as an area under the P-V curve.

For example, one part of the work done by the diaphragm on the lungs due to inspiration can be thought of as the work needed to overcome the elastic resisting forces of the chest wall and diaphragm. This work done to overcome elasticity can be represented by the area under the line AB as shown in Fig. 3.5.

The total work done by the diaphragm on the lungs due to inspiration can be thought of as the work needed to overcome the elastic resisting forces of the chest wall and diaphragm, plus the work done to

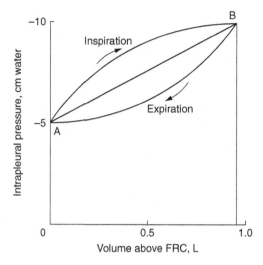

Figure 3.4 Intrapleural pressure versus lung volume for inspiration and expiration.

overcome the resistance to flow. This total work of inspiration can be represented by the area under the inspiration curve as shown in Fig. 3.6.

During expiration, the elasticity of the lungs and chest wall helps provide the stored energy so that diaphragm work is not necessary. The energy stored in these elastic tissues can be partially recovered during expiration. The work done to overcome resistance to flow cannot be overcome. The total work of one total breathing cycle, including inspiration and expiration, can be shown as the cross-hatched area in Fig. 3.7.

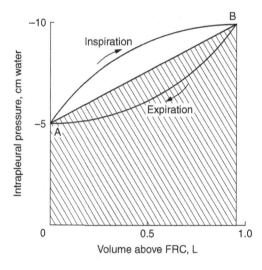

Figure 3.5 Pressure-volume curve showing the work done to overcome elasticity during inspiration.

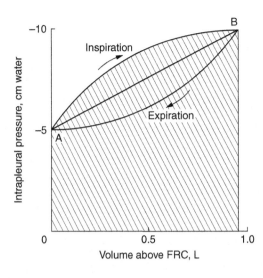

Figure 3.6 Pressure-volume curve showing the total work done by diaphragm due to inspiration.

3.5.1 Clinical features: respiratory failure

An important cause of respiratory failure is fatigue of the muscles of respiration. When the diaphragm and respiratory muscles cannot carry out the work of breathing, the result is a progressive fall in oxygenation and/or rise in carbon dioxide concentration.

(Lingappa, 2000)

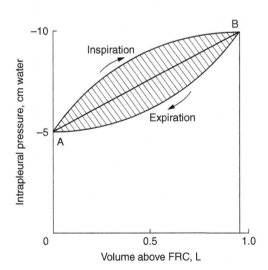

Figure 3.7 Work done by the diaphragm during one breath cycle.

3.6 Airway Resistance

Although the air that flows through your trachea is not very massive or very viscous, there is a noticeable hydraulic resistance to the flow. This resistance results in a pressure drop along the airway. This pressure decreases along the airways in the direction of flow. This pressure drop is also dependent on the flow rate in the tube, the viscosity of the fluid, and the pattern of flow. There can be no flow along a tube unless there is a pressure difference, or pressure gradient, along the tube.

As explained in Sec. 1.4.1, fluid particles move along streamlines during laminar flow. When air flows at low rates in relatively small diameter tubes, as in the terminal bronchioles, the flow is laminar. Turbulent flow is a random mixing flow. When air flows at higher rates in larger diameter tubes, like the trachea, the flow is often turbulent.

In Chap. 1, we also saw that a dimensionless parameter, termed the Reynolds number, Re, could be used to predict whether flow is turbulent or laminar. The number is defined in Eq. (3.7):

$$\text{Re} = \frac{\rho VD}{\mu} \tag{3.7}$$

In Eq. (3.7) ρ is fluid density in kg/m^3, V is fluid velocity in m/s, D is pipe diameter in m, and μ is fluid viscosity in Ns/m^2. Physically, the Reynolds number represents the ratio of inertial forces to viscous forces.

As an analogy, imagine the ratio of the momentum of a vehicle (mass × velocity) to the frictional braking force available (force). A person walking on dry pavement has a relatively small mass, relatively low velocity, and relatively low ratio of momentum to stopping force. For comparison, imagine a very large truck moving at high velocity on an icy street. This second combination of truck on ice has a very high ratio of momentum to stopping force and is a highly unstable situation in comparison to the first. Analogously to the person walking on pavement, low-mass airflows with low Reynolds numbers are more stable and more likely to be laminar in comparison to denser flows at high velocity with higher Reynolds numbers.

In the lungs, fully developed laminar flow probably occurs only in very small airways with low Reynolds number. Flow in the trachea may be truly turbulent. Much of the flow in intermediate-sized airways will be transitional flow in which it is difficult to predict whether the flow will be laminar or turbulent.

In Chap. 1, we also developed Poiseuille's law which describes laminar flow in rigid tubes. Poiseuille's law applies to airflow, just as it does to blood flow, when the flow is laminar. Recall that:

$$Q = \frac{-\pi R^4}{8\mu} \frac{dP}{dx} \tag{3.8}$$

where Q is the flow rate, R is the airway radius, μ is the air viscosity and dP/dx is the pressure gradient along the airway. In the same way that electrical resistance is the ratio of driving voltage to current flow, resistance to flow can be thought of as the driving pressure divided by the volumetric flow rate. Poiseuille's law can be rearranged to solve for the resistance as shown in Eq. (3.9).

$$\text{Resistance} = \frac{8\mu}{\pi R^4} \tag{3.9}$$

Although the resistance to flow associated with the tube is inversely related to the fourth power of the radius, the resistance contributed by airways is not predominantly in the smallest diameter airways. As airways branch they become narrower, but also more numerous. The major site of airway resistance is the medium-sized bronchi. Small bronchioles contribute relatively little resistance because of their increased numbers.

Most resistance in airways occurs up to the seventh generation of branching vessels. Less than 20 percent of the resistance is attributable to airways that are less than 2 mm in diameter because of the large number of vessels. Between 25 and 40 percent of total resistance is in the upper airways including the mouth, nose, pharynx, larynx, and trachea.

One way to assess expiratory resistance is to begin by measuring the forced vital capacity of the lung, FVC, using a spirometer. The patient starts by making a maximal inspiratory effort to TLC. After a short pause, the patient makes a maximal forced expiratory effort.

We may define forced expiratory volume, FEV_1, as the volume of air expired in 1 s during a forced expiratory effort. The ratio of FEV_1 to FVC is a good index of airway resistance. In normal, healthy subjects, FEV_1/FVC is greater than 0.80, or 80 percent.

Forced expiratory flow rate between time t_1 and time t_2 can be defined as the average flow rate between times t_1 and t_2, as measured during a forced expiration. The rate can be represented graphically as the slope of a line drawn between two points on the forced expiration curve.

In Fig. 3.8, point A represents the point when the lung volume is equal to 25 percent of the VC. A second point, point B, is drawn on the curve representing the point when lung volume is equal to 75 percent of VC. The slope of the line AB is the forced expiratory flow rate, $FEF_{25-75\%}$.

Figure 3.8 represents a normal, healthy patient and Fig. 3.9 represents a patient with an airway obstruction. Note the steep slope corresponding to a high forced expiratory flow rate in the healthy patient in Fig. 3.8 and the shallow slope corresponding to a low forced expiratory flow rate in Fig. 3.9.

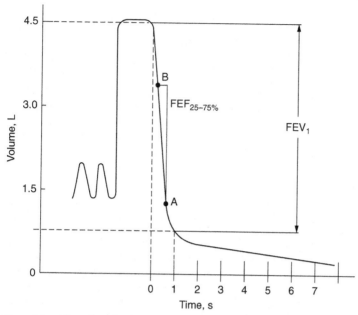

Figure 3.8 A forced expiration versus time curve for a patient with normal airway resistance.

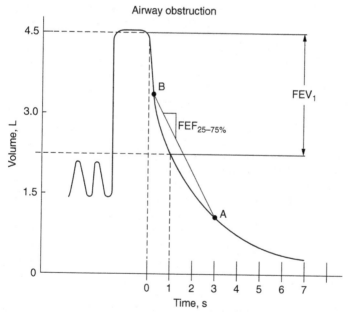

Figure 3.9 The volume versus time curve for forced expiration in a patient with chronic obstructive pulmonary disease shows a great resistance to expiratory flow.

3.6.1   Example problem: Reynolds number

Two pipelines have equal diameters. Water flows in one pipeline and air flows in the second. Temperature is constant and equal in both pipelines. You can control the average flow velocity which remains equal in both pipelines. As you increase velocity in both pipes, which flow will become turbulent first, the water or the air?

First, we need some values for density and viscosity or kinematic viscosity for both fluids. We'll use the following values.

Density   air-1.23 kg/m^3
                                  Water-999 kg/m^3

Viscosity   air-1.78 × 10^{-5} Ns/m^2
                                  Water-1.12 × 10^{-3} Ns/m^2

Kinematic Viscosity   air-1.46 × 10^{-5} m^2/s
                                  Water-1.12 × 10^{-6} m^2/s

Since the velocity and pipe diameters will be equal in both pipelines, we can look at the Reynolds number divided by the velocity and pipe diameter, which is the reciprocal of kinematic viscosity.

$$\frac{R_E}{VD} = \frac{\rho}{\mu} = \frac{1}{\text{kinematic viscosity}}$$

Now we can compare the reciprocal of the kinematic viscosity for each fluid. The number with the largest value (smallest kinematic viscosity) will become turbulent first. Water has a much smaller kinematic viscosity. **The water pipeline becomes turbulent first. The ratio of Reynolds numbers is 13:1 with the higher Reynolds number in the water pipeline. Although the viscosity of air is small, the density of the air is so low that the airflow represents a more stable situation.**

3.7   Gas Exchange and Transport

In the next section of this book, we will consider how oxygen moves from ambient air into the tissues of the body. Diffusion of a gas occurs when there is a net movement of molecules from an area with a high partial pressure to an area with a lower partial pressure. Only 50 years ago, it was still believed by some scientists that the lung secreted oxygen into the capillaries. This would mean that oxygen would move from the atmosphere to a relatively higher concentration inside the lung by an active process. More accurate measurements have now shown that gas transport across the alveolar wall is a passive process.

3.7.1 Diffusion

Gases like oxygen and carbon dioxide move across the blood-gas barrier of the alveolar wall by diffusion. You might ask, "what parameters affect the rate of transfer?" The rate of gas movement across the alveolar wall is dependent on the diffusion area, the driving pressure, and the wall thickness.

The diffusion area is the surface area of the **alveolar wall** or **blood-gas barrier**. That surface area is proportional to the rate of diffusion. The driving pressure which pushes gases across the alveolar wall is the partial pressure of the gas in question. The driving force that pushes oxygen into the bloodstream is the difference between the partial pressure of oxygen in the alveoli and the partial pressure of oxygen in the blood, ΔP_{O_2}. The rate of diffusion of oxygen into bloodstream is proportional to ΔP_{O_2}.

The rate of diffusion of oxygen into the bloodstream is inversely proportional to the thickness of the alveolar wall. A thicker wall causes the oxygen diffusion to decrease and a thinner wall makes it easier for oxygen to flow into the bloodstream.

The blood-gas barrier in the lung has a surface area of about 50 to 100 m^2 shared by over 750 million alveoli. This huge diffusion area is available to a relatively small volume of blood. Pulmonary capillary blood volume is only about 60 mL during resting and about 95 mL during exercise.

Carbon dioxide diffuses about 20 times more rapidly than oxygen through the alveolar wall. The much higher solubility of carbon dioxide is responsible for this increased diffusion rate.

3.7.2 Diffusing capacity

One might wonder if oxygen transfer into the blood is limited by how fast the blood can flow through the lungs, or by how fast oxygen can diffuse through the blood-gas barrier. It turns out that under normal physiological circumstances, oxygen diffusion through the alveolar wall is sufficient to provide oxygen to all red blood cells flowing through the pulmonary capillaries, and the limiting factor for oxygen uptake is perfusion or the rate of blood flow through those capillaries. A red blood cell, also referred to as an erythrocyte, spends an average of about 0.75 to 1.2 s passing through a pulmonary capillary under normal resting conditions.

Under some circumstances oxygen diffusion rates through the wall may also limit the oxygen transfer rate. During extreme exercise, the erythrocyte may stay in the capillary as little as 0.25 s. Even this very short time would be enough time for oxygen to diffuse into the capillaries at normal atmospheric oxygen partial pressures. However, during extreme exercise at high altitude, or in a patient with thickening of the alveolar wall due to pulmonary fibrosis, oxygen flow rate may switch from a perfusion-limited to a diffusion-limited process.

Resistance to diffusion So far, we have considered the blood-gas barrier as the only source of resistance to diffusion of gases from the alveoli into the bloodstream. In fact, there is a second important component. The uptake of oxygen occurs in two stages. The first is diffusion through the alveolar wall and the second is the reaction of oxygen with **hemoglobin**.

The diffusing capacity through the alveolar wall is defined by the flow rate of the gas divided by the partial pressure difference that is driving the flow. **Diffusing capacity** of a gas can be written as D_a in Eq. (3.10).

$$D_a = \frac{\dot{V}_{\text{gas}}}{(P_a - P_c)} \tag{3.10}$$

In Eq. (3.10), $\dot{V}_{\text{gas}}$ represents the flow rate of some gas from the alveoli into the capillary. P_a represents the partial pressure of that gas in the alveoli and P_c represents the partial pressure of the same gas in the capillary.

If we think of the electrical analogy in which resistance is equal to voltage divided by current, we can see that it is possible to think of $1/D_a$ as a **resistance to diffusion**.

$$\frac{1}{D_a} = \frac{P_a - P_c}{\dot{V}_{\text{gas}}} \tag{3.11}$$

The rate of reaction of oxygen with hemoglobin can be represented by the symbol θ. The units on θ are $\frac{\text{mL O}_2}{\text{min}}$ per mL blood per mmHg partial pressure of O_2. If V_c represents the volume of pulmonary capillary blood, then the product of θ and V_c will have the units of mL O_2/min/mmHg or flow divided by pressure. Once again, the inverse, pressure/flow is analogous to resistance.

$$\text{Resistance}_{O_2\text{-Hgb}} = \frac{1}{\theta V_c} \tag{3.12}$$

The total resistance to the flow of oxygen into the bloodstream is the combination of the resistance to diffusion caused by the blood-gas barrier plus the resistance to flow due to the oxygen-hemoglobin reaction.

The resistances can be added together as shown in Eq. (3.13). The quantity $1/D_L$ is the total resistance to the flow of oxygen due to the lung.

$$\frac{1}{D_L} = \frac{1}{D_a} + \frac{1}{\theta V_c} \tag{3.13}$$

For oxygen flow, the resistance to diffusion offered by the membrane is approximately equal to the resistance associated with the oxygen-hemoglobin reaction. It is also interesting to note that carbon dioxide diffusion is approximately twenty times faster than oxygen diffusion. Therefore, it seems unlikely that CO_2 elimination will be slowed by an increased resistance to diffusion.

3.7.3 Oxygen dissociation curve

One gram of pure hemoglobin can combine with 1.34 mL of oxygen, and normal human blood has approximately 15 g of hemoglobin in each 100 mL of whole blood. This means that the oxygen-carrying capacity of blood is about 20.8 mL oxygen/100 mL blood.

Oxygen saturation is reported as a percentage and is equal to the amount of oxygen combined with hemoglobin at a given partial pressure divided by the maximum oxygen-carrying capacity. For example, oxygen saturation of blood is about 75 percent when the blood is exposed to air with a partial pressure of oxygen of 40 mmHg and 97.5 percent saturation at a partial pressure of oxygen of 100 mmHg. Figure 3.10 shows a normal oxygen dissociation curve from a human.

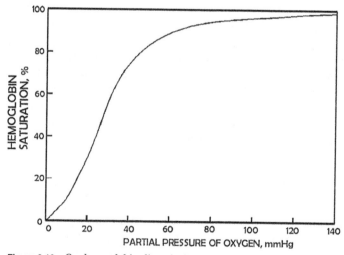

Figure 3.10 Oxyhemoglobin dissociation curve.

Oxygen dissolved in blood, rather than carried in the hemoglobin, amounts to 0.003 mL O_2 per 100 mL blood per mmHg partial pressure. Therefore, the amount of dissolved oxygen in blood with a driving pressure of oxygen equal to 100 mmHg is only 0.3 mL O_2 per 100 mL blood, compared to 20.3 mL O_2 carried in the hemoglobin.

To make the connection between oxygen content in blood (mL O_2 per 100 mL blood) and oxygen partial pressure (mmHg), we can use the **oxygen content equation.**

$$Ca_{O_2} = \left(Sa_{O_2} \times \text{grams of Hemoglobin} \times 1.34 \, \frac{\text{mL } O_2}{\text{gram Hgb}} \right) + (0.003 \, Pa_{O_2})$$

where Ca_{O_2} is the oxygen content of arterial blood in $\frac{\text{mL } O_2}{100 \text{ mL blood}}$, Sa_{O_2} is oxygen saturation in percent, and Hgb is hemoglobin content in the blood in g/100 mL blood. The oxygen-carrying capacity of 1 gram of hemoglobin is 1.34 mL. The term farthest to the right in the oxygen content equation represents the small amount of oxygen that is dissolved in plasma.

3.7.4   Example problem: oxygen content

A lowlander has 15 g hemoglobin (Hgb) per 100 mL blood and an oxygen saturation of 98 percent. Based on the hemoglobin saturation curve shown below, how much hemoglobin would a high-altitude native need in order to have the same amount of oxygen in his/her blood at 4500 m above mean sea level? The barometric pressure at 4500 m is 430 mmHg. Take the alveolar concentration of oxygen to be 50 mmHg. The chart below shows hemoglobin saturation versus alveolar partial pressure of oxygen. Ignore the small amount of dissolved oxygen in the plasma.

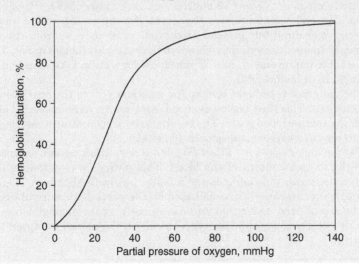

Solution The oxygen saturation in the lowlander is 85 percent from the above chart.

The lowlander with 98 percent hemoglobin saturation will have an oxygen content of his/her blood of:

$$1.34 \, \frac{\text{mL O}_2}{\text{gm Hgb}} \left(15 \, \frac{\text{gm Hgb}}{100 \text{ mL blood}} \right)(0.98) = 19.7 \, \frac{\text{mL O}_2}{100 \text{ mL blood}}$$

To find the oxygen content of the blood in the high-altitude native solve for the hgb content that will give the same total oxygen content at 85 percent saturation:

$$20.4 \, \frac{\text{mL O}_2}{100 \text{ mL blood}} = 1.34 \, \frac{\text{mL O}_2}{\text{gm Hgb}} \left(X \, \frac{\text{gm Hgb}}{100 \text{ mL blood}} \right)(0.85)$$

$$X = 17.9 \, \frac{\text{gm Hgb}}{100 \text{ mL blood}}$$

corresponding to about a 19 percent increase in hematocrit or an increase of hematocrit from 0.42 to 0.50. Richalet et al. did a study on Chilean miners who commute from sea level to 4500 m and found that after a year of chronic intermittent hypoxia (7 days at high altitude followed by 7 days at sea level), the mean hematocrit for the miners increased to 52 percent (Richalet et al., 2002).

3.7.5 Clinical feature

A 54-year-old man was admitted to the emergency room complaining of headache and shortness of breath. On room air, his partial pressure of arterial oxygen (Pa_{O_2}) was 89 mmHg, partial pressure of CO_2 (Pa_{CO_2}) was 38 mmHg, and hematocrit was 44 percent. His oxygen saturation was not directly measured but instead calculated at 98 percent from the oxyhemoglobin dissociation curve shown in the example problem in Sec. 3.7.4. After some improvement, he was scheduled for a brain CAT scan and discharged from the hospital.

The patient was brought back to the emergency room the next evening, unconscious. This time carbon monoxide and oxygen were measured along with the routine blood gases. The results were oxygen saturation (Sa_{O_2}) = 53 percent and carboxyhemoglobin 46 percent.

Carbon dioxide does not affect Pa_{O_2} but it does affect oxygen saturation and the oxygen content of the blood. This patient was suffering from carbon monoxide poisoning due to a faulty heater in his home. The oxyhemoglobin saturation curve is based on the partial pressure of oxygen in the alveoli, and use of that curve assumes that the hemoglobin in the arteries comes to equilibrium with the air in the alveoli (adapted from Martin).

3.8 Pulmonary Pathophysiology

3.8.1 Bronchitis

Bronchitis is an inflammation of the airways resulting in excessive mucus production in the bronchial tree. Bronchitis occurs when the inner walls of the bronchi become inflamed. It often follows a cold or other respiratory infection and happens in virtually all people, just as the common cold. When the bronchitis does not go away quickly but persists, then it is termed chronic bronchitis.

3.8.2 Emphysema

Emphysema is a chronic disease in which air spaces beyond bronchioles are increased. The stiffness of the alveoli is decreased, (static compliance is increased), and airways collapse more easily. Because of the decreased stiffness of the lung, exhalation requires active work and the work of breathing is significantly increased. The surface area of the alveoli become smaller, and the air sacs become less elastic. As carbon dioxide accumulates in the lungs, there becomes less and less room available for oxygen to be inhaled, thereby decreasing the partial pressure of oxygen in the lungs. See Fig. 3.11 which shows a picture of an emphysematous lung.

Emphysema is most often caused by cigarette smoking, although some genetic diseases can cause similar damage to the alveoli. Once this damage has occurred, it is not reversible.

3.8.3 Asthma

Asthma is a chronic disease which affects 5 million children in the United States. In asthma, the airways become overreactive with increased mucus production, swelling, and muscle contraction. Because of the decreased size of the bronchi and bronchioles, flow of air is restricted and both inspiration and expiration become more difficult.

3.8.4 Pulmonary fibrosis

Pulmonary fibrosis affects 5 million people worldwide and 200,000 in the United States. Pulmonary fibrosis is caused by a thickening or scarring of pulmonary membrane. The result is that the alveoli are gradually replaced by fibrotic tissue becoming thicker, with a decreased compliance (increased stiffness) and a decrease in diffusing capacity. Symptoms of pulmonary fibrosis include a shortness of breath, chronic dry, hacking cough, fatigue and weakness, chest discomfort, loss of appetite, and rapid weight loss. Traditionally, it was thought that pulmonary fibrosis might be an autoimmune disorder or the result of a viral infection. There is growing evidence that there is a genetic link to pulmonary fibrosis.

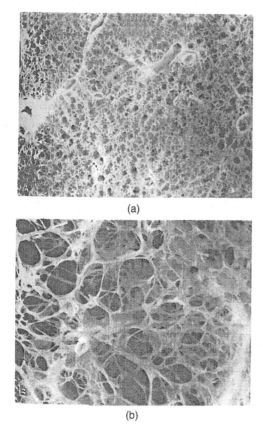

(a)

(b)

Figure 3.11   Normal lung versus an emphysematous lung. (a) Normal lung; (b) emphysematous lung. (*With permission from West, 1992, Figure 4.3, p. 61.*)

3.8.5   Chronic obstructive pulmonary disease (COPD)

Chronic obstructive pulmonary disease (COPD) is a slowly progressive disease of the lung and airways. COPD can include asthma, chronic bronchitis, chronic emphysema, or some combination of these conditions. The disease is characterized by a gradual loss of lung function. The most significant risk factor for COPD is cigarette smoking. Other documented causes of COPD include occupational dusts and chemicals. Genetic factors can also play a significant role in some forms of this disease.

3.8.6   Heart disease

Heart disease should be mentioned in any discussion of pulmonary pathologies. While cardiac disease is not strictly speaking, a pulmonary pathology, some forms of cardiac disease can certainly lead to respiratory pathologies. For example, a stenotic regurgitant mitral valve can cause back pressure in pulmonary capillaries leading to fluid in the lungs.

3.8.7   Comparison of pulmonary pathologies

Figure 3.12 shows the volume versus time curve for a normal lung compared to that for a patient with fibrosis, asthma, and emphysema. In the patients with fibrosis, asthma, and emphysema, note the shallow slope

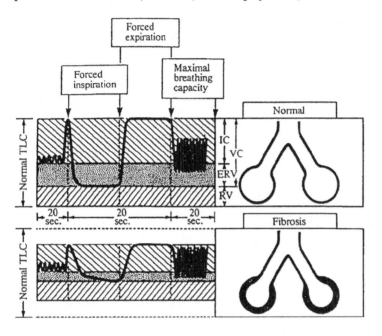

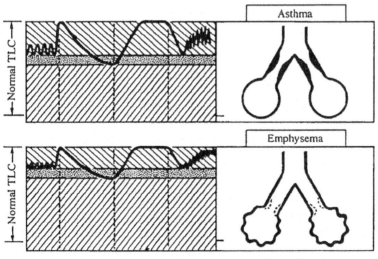

Figure 3.12   Spirometer comparisons between a normal lung, fibrosis, asthma, and emphysema. (*From Selkurt, 1982, Figure 9.20, p. 342.*)

of the curve during forced inspiration and expiration. In other words, the change in volume over time, dV/dt, is much smaller in patients with lung disease compared to the dV/dt in the normal lung. The airflow rate is much smaller in all three cases.

3.9   Respiration in Extreme Environments

Consider how you might feel if you drive your automobile to the top of Pikes Peak (14,109 ft above sea level, ASL) or if you ride a cable car to the top of the Zugspitze, the highest point in Germany (9718 ft ASL). If you have had the opportunity to visit either of these locations, you probably experienced the shortness of breath associated with breathing in environments with low oxygen pressure. (The percentage of oxygen does not vary much with the increase in altitude, but the partial pressure of oxygen diminishes.) Perhaps you even developed a headache after a short period. How you felt was dependent on how long it took to achieve the altitude, how long you remained, how well hydrated you may have been at the time, and a number of other potential factors.

3.9.1   Barometric pressure

Just as with normal respiration, at high altitude the driving force which helps to push oxygen into your blood is the partial pressure of oxygen. This partial pressure depends on both the barometric pressure and the relative percentage of air that consists of oxygen. Barometric pressure depends on the altitude above the earth's surface and varies approximately exponentially as shown in Fig. 3.13.

The equation for barometric pressure as a function of altitude depends on the density of the air at varying altitudes and therefore on air temperature. The equation for the standard atmosphere between sea level and 11 km above the earth's surface can be given by:

$$P_{atm} = 760(1 - 0.022558z)^{5.2559}$$

where P_{atm} is the barometric pressure in millimeters of mercury, and z is the altitude above mean sea level in kilometers.

From Fig. 3.13, it is possible to see that on a standard day, the barometric pressure would decrease from 747 mmHg in Terre Haute, Indiana, to 625 mmHg in Denver, Colorado, to 360 mmHg at the summit of Mt. Kilimanjaro, and 235 mmHg at the summit of Mt. Everest. With the decrease in altitude and corresponding decrease in barometric pressure comes a decrease in the partial pressure of oxygen.

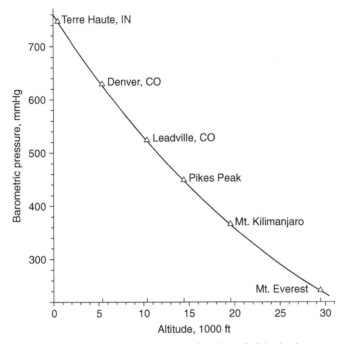

Figure 3.13   Barometric pressures as a function of altitude above sea level.

3.9.2 Partial pressure of oxygen

The partial pressure of oxygen, P_{O_2} is the driving pressure for getting oxygen into the blood. Air is 21 percent oxygen, so the partial pressure of oxygen in standard air is 0.21 times the barometric pressure. This is also known as the fraction of inspired oxygen, $F_{I_{O_2}}$. The P_{O_2} in air on a standard day in Terre Haute, Indiana, is $0.21 \times 747 = 157$ mmHg. If P_{O_2} is low, the driving pressure to push oxygen into the bloodstream will be low, making it more difficult to breathe.

Further, the air inside your lungs is not dry air. Water vapor also displaces oxygen. In fact, the air in your lungs is saturated with water, and the vapor pressure of water at 37°C, the temperature inside your lungs, is 47 mmHg. Now the P_{O_2} of sea-level dry air is

$$760 \,(.2093) = 159 \text{ mmHg,}$$

but the P_{O_2} of saturated, inspired air is

$$(760 - 47)(.2093) = 149 \text{ mmHg.}$$

If you climb even higher, the driving force becomes lower but the vapor pressure of water doesn't change since the air in your lungs is always saturated (100 percent relative humidity).

Let's climb to 14,000 ft above sea level (ASL). Now the P_{O_2} of inspired air at 14,000 ft ASL is (446 mmHg − 47)(.2093) = 82 mmHg or about half of the P_{O_2} for inspired air at sea level.

Now if we continue to climb to 18,000 ft ASL, the P_{O_2} in dry air is half that of sea-level dry air and the inspired partial pressure of oxygen is only 70 mmHg.

$$(380 - 47)\,(.2093) = 70 \text{ mmHg}$$

Finally, at the top of Mt. Everest the P_{O_2} is one-third of sea-level dry air and the P_{O_2} of inspired air at 29,000 ft ASL is only 43 mmHg.

$$(250 - 47)\,(.2093) = 43 \text{ mmHg}$$

What would happen if a high performance jet lost cabin pressure at an altitude of 63,000 ft ASL? The barometric pressure at this extreme altitude is less than the vapor pressure of water at 37°C. In this case the partial pressure of oxygen would be zero and, in fact, your blood (and all other water in your tissues) would boil.

3.9.3 Hyperventilation and the alveolar gas equation

Inspired oxygen is not the complete story. What about the CO_2 in your lungs? Doesn't it also displace air, making the P_{O_2} even lower? The amount of oxygen in the alveoli depends on the production of carbon dioxide and the rate of transfer of oxygen from the lungs to the blood in the pulmonary capillaries.

To understand what is happening with the person's arterial oxygen concentration, begin with the alveolar gas equation (Levitzky, 2003) which takes into account the patient's arterial CO_2 partial pressure (Pa_{CO_2}), fraction of inspired oxygen ($F_{I_{O_2}}$), and the barometric pressure (P_B). Take care with the symbols since "a" in Pa_{CO_2} represents "arterial" while "A" in $P_{A_{CO_2}}$ stands for "alveolar."

$$P_{A_{O_2}} = F_{I_{O_2}}(P_B - P_{H_2O}) - P_{A_{CO_2}}\left(F_{I_{O_2}} + \frac{1 - F_{I_{O_2}}}{R}\right)$$

where P is partial pressure, A is alveolar, $F_{I_{O_2}}$ is fraction of inspired oxygen, P_B is barometric pressure, P_{H_2O} is the vapor pressure of water at 37°C, and R is the respiratory quotient.

The respiratory quotient or respiratory exchange ratio is:

$$R = \frac{\dot{V}_{CO_2}}{\dot{V}_{O_2}} = \frac{\text{rate of } CO_2 \text{ production}}{\text{rate of } O_2 \text{ usage}}$$

This alveolar gas equation is valid if there is no CO_2 in inspired gas. If we assume a typical value for R of 0.8, the following abbreviated alveolar gas equation is often used for clinical purposes.

$$P_{A_{O_2}} = F_{I_{O_2}}(P_B - 47) - Pa_{CO_2}(1.2)$$

This equation uses a water vapor pressure of 47 mmHg, which is the vapor pressure of water at body temperature, 37°C. Ambient $F_{I_{O_2}}$ is the same at all altitudes, 0.21.

The partial pressure of carbon dioxide in your lungs can approach 40 mmHg. Since the partial pressure of inspired oxygen at an altitude of 18,000 ft ASL was calculated as 70 mmHg without considering CO_2, the P_{O_2} of air in the alveoli could be as low as $70 - 1.2(40) = 22$ mmHg. At the top of Mt. Everest it could be as low as $43 - 1.2(40) = -5$ mmHg. If that is true, how can so many people climb above 14,000 ft so easily and a few people even reach the summit of Mt. Everest without oxygen? The short answer is hyperventilation. By breathing faster, climbers are able to lower the partial pressure of carbon dioxide in their alveoli. If you increase ventilation rate by four, you can lower the P_{CO_2} to about 10 mmHg.

By hyperventilation, the P_{O_2} of alveolar oxygen at the top of Mt. Everest can be calculated as follows.

P_{O_2} of inspired air at 29,000 ft is $(250 - 47)(.2093) = 43$ mmHg. The P_{CO_2} of the air in the alveoli is 10 mmHg. The partial pressure of the oxygen in the alveoli is $43 - 1.2(10) = 30$ mmHg.

3.9.4   Alkalosis

One might wonder whether there are any difficulties or side effects resulting from hyperventilation and the associated drop in the partial pressure of carbon dioxide. In fact, the result is respiratory alkalosis. The pH of the blood increases above normal. You feel bad and you can't sleep well. One result could be acute mountain sickness.

Your body's solution to respiratory alkalosis is that your kidneys will excrete bicarbonate over the next few days. At the same time, the increase in pH puts a kind of "brake" on ventilation and causes your breathing to slow down. The excretion of bicarbonate causes the pH in the blood to decrease and the "brakes" on ventilation are reduced. After two or three days at altitude, blood pH returns to normal.

3.9.5   Acute mountain sickness

The feeling of nausea and headache, resulting from low oxygen environments, hyperventilation, and the associated respiratory alkalosis, is known as acute mountain sickness (AMS). Sleeping at altitude is also troublesome. When your breathing slows, oxygen saturation in your blood drops.

This drop in P_{O_2} causes you to wake up with a feeling of breathlessness. Diamox is a trademark name for acetazolamide, a well-known carbonic anhydrase inhibitor which is used in the treatment of seizures and glaucoma and is also a diuretic. Diamox does not prevent mountain sickness, but it speeds up acclimatization by increasing urinary excretion of bicarbonate. People taking Diamox have been shown to have a more consistent level of blood oxygen saturation, enabling more restful sleep. AMS can also result in nocturnal periodic breathing, weird dreams, and frequent awakening at night.

3.9.6 High-altitude pulmonary edema

High-altitude pulmonary edema (HAPE) is a life-threatening noncardiogenic (not caused by heart disease) lung edema. The mechanism that causes HAPE is not completely known, but the disease is thought to be caused by patchy, low-oxygen, pulmonary vasoconstriction (constriction of blood vessels). That constriction results in localized overperfusion and increased permeability of pulmonary capillary walls. These changes result in high pulmonary artery pressure, high permeability, and fluid leakage into the alveoli. The result of HAPE is that the lungs fill with fluid, decreasing ventilation. Symptoms of HAPE can include cough, shortness of breath on exercise, progressive shortness of breath, and eventually suffocation, if left untreated. The condition is unstable and the only effective treatment for HAPE is descent to a lower altitude and respiration in a more oxygen-rich environment.

3.9.7 High-altitude cerebral edema

High-altitude cerebral edema (HACE) is a less common, but equally life-threatening condition in which cerebral edema is the result of breathing in a depleted oxygen environment. HACE is theoretically linked to brain swelling. Symptoms of HACE can include severe throbbing headache, confusion, difficulty walking, difficulty speaking, drowsiness, nausea, vomiting, seizures, hallucinations, and coma. A person suffering from HACE is often gray or pale in appearance. A victim can suffer from HAPE and HACE simultaneously. Both HAPE and HACE normally occur at altitudes above 15,000 ft ASL but can occur at high ski areas above 8000 ft ASL.

3.9.8 Acclimatization

If you ascend to a high altitude and remain for a long period of time, your body can begin to acclimatize. By increasing the number of red blood cells in your blood, you are able to increase the amount of hemoglobin and hence the oxygen carrying capacity of your blood. There are people who live permanently above 16,000 ft in the Peruvian Andes. Their inspired P_{O_2} is only about 45 mmHg, but those people have more oxygen in their blood than do normal sea-level residents!

The volume percent of red blood cells in blood is known as hematocrit. Hematocrit in normal healthy males is about 42 to 45 percent. If a person's hematocrit falls below 25 percent, then he or she is considered anemic. Hematocrit values can also be greater than normal and a hematocrit above 70 percent is known as polycythemia.

During acclimatization to low oxygen environments, it is possible to replace approximately 1 percent of your erythrocytes per day. At the same time, erythrocytes continue to die after their approximately 125-day life span. Therefore, it takes 2 or 3 weeks to increase your hematocrit significantly.

In a study done in Potosi, Bolivia, which is located in southern Bolivia on a high plane at approximately 13,000 ft ASL, the average hematocrit in males was found to be 52.7 percent. Early observations of human adaptation to high altitude were made in Potosi by de la Calancha in the late 1600s. The Spanish historian writes that "more than one generation was required" for a Spanish child to thrive in Potosi. It's not clear whether being born and raised at high altitude or whether interbreeding with the Andean population was required.

3.9.9 Drugs stimulating red blood cell production

Erythropoietin (EPO) is a naturally occurring hormone which stimulates red blood cell production. Low oxygen stimulates EPO production in humans, which in turn stimulates a higher production of red blood cells. Recombinant EPO (rEPO) is a synthetic version of this hormone.

The difficulty with taking rEPO to increase hematocrit is that an increase in hematocrit also increases blood viscosity, and therefore increases the work of the heart required to pump blood. At the 1984 Los Angeles Olympics before rEPO was available, some American bicycle racers received blood transfusions to raise their red blood cell counts (they also received hepatitis). After rEPO became readily available, athletes began taking injections of rEPO. A number of Belgian and Dutch professional cyclists died of strokes in 1987–88, presumably from EPO-induced clots in arteries.

Recombinant human erythropoietin (rEPO) has become the standard of care for renal anemia. rEPO has a relatively short half-life and is generally administered two or three times a week.

Darbepoetin is a synthetic hormone that increases red blood cell production, and is used to treat anemia and related conditions. It is in the same class of drugs as rEPO and competes for the same market. Its brand name is Aranesp, and it is marketed by Amgen. It was approved in September 2001 by the U.S. Food and Drug Administration for treatment of patients with chronic renal failure by intravenous or subcutaneous injection. Darbepoetin alpha is a longer lasting agent with a half-life approximately three times that of rEPO.

Like rEPO, Darbepoetin's use increases the risk of cardiovascular problems, including cardiac arrest, seizures, arrhythmia or strokes, hypertension, congestive heart failure, vascular thrombosis, myocardial infarction, and edema. Also like rEPO, Darbepoetin has the potential to be abused by athletes seeking an advantage. Its use during the 2002 Winter Olympic Games to improve performance led to the disqualification of several cross-country skiers from their final races.

3.9.10   Example problem: alveolar gas equation

A high-altitude native living in the Andes in Potosi, Bolivia, at 13,000 ft (~4000 m) above sea level has a hematocrit 53 percent. If the partial pressure of alveolar oxygen in this man's lungs is 50 mmHg, calculate the partial pressure of arterial carbon dioxide using the abbreviated form of the alveolar gas equation.

Solution   The abbreviated form of the alveolar gas equation is:

$$PA_{O_2} = F_{I_{O_2}}(P_B - 47) - 1.2(Pa_{CO_2})$$

Solving for the partial pressure of CO_2 yields:

$$Pa_{CO_2} = \frac{F_{I_{O_2}}(P_B - 47) - Pa_{O_2}}{1.2}$$

To solve for the barometric pressure at altitude, from Sec. 3.9.1, we use the equation:

$$P_{atm} = 760(1 - 0.022558z)^{5.2559}$$

$$P_{atm} = 760(1 - 0.022558(4))^{5.2559} = 462 \text{ mmHg}$$

Therefore,
$$Pa_{CO_2} = \frac{0.21(462 - 47) - 50}{1.2} = 31 \text{ mmHg}$$

The partial pressure of carbon dioxide in the arterial blood is 31 mmHg.

Review Problems

1. Determine the total lung capacity of a patient with vital capacity of 4.6 L, functional residual capacity of 2.6 L, and residual volume of 1.3 L.

2. Determine the expiratory reserve capacity of a patient with functional residual capacity of 2.9 L and a residual volume of 1.3 L.

3. Find the vital capacity for a person with a total lung capacity of 7 L and a residual volume of 1.5 L.

4. Find the inspiratory reserve volume for a patient with a total lung capacity of 7 L, a functional residual capacity of 2.7 L, and a tidal volume of 500 mL.

5. Find the rate of breathing in breaths per minute for a patient with a tidal volume of 500 mL, a cardiac output of 5 L/min, and a ventilation/perfusion ratio of 1.1.

6. A person has a breathing rate of 18 breaths per minute with a tidal volume of 400 mL, a heart rate of 100 bpm, and a stroke volume of 90 mL. What is this person's ventilation to perfusion ratio.

7. Find the average compliance of the lung, in m^5/N, during tidal breathing for the patient from whom the static pressure-volume diagram shown below is generated.

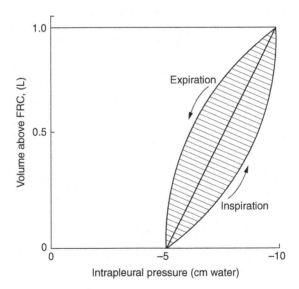

8. Calculate the work of breathing, in Joules, for a single breath for a patient who generated the static pressure-volume diagram shown below. If the person is taking 12 breaths per minute, what is the power expended by breathing? The cross-hatched area is 1.91 L · cm H_2O:

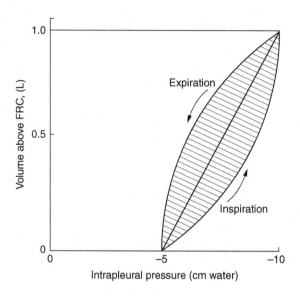

9. The figure below shows a volume-time curve for a patient with chronic obstructive pulmonary disease (COPD). Find the forced expiratory flow rate, $FEF_{25-75\%}$, for this patient. Assume the residual volume, RV, to be 0.5 L.

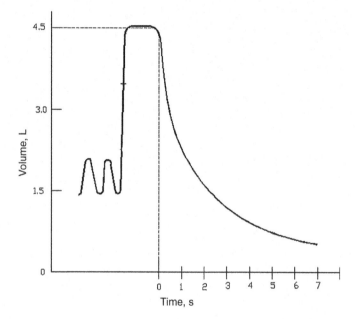

10. In normal healthy subjects the ratio of forced expiratory volume to force vital capacity, FEV_1/FVC ratio, is greater than:

(a) 20%
(b) 40%
(c) 60%
(d) 80%
(e) 100%

11. Estimate the atmospheric pressure at 14,000 ft above mean sea level.

12. A given person has 15 g Hgb per 100 mL blood and a partial pressure of alveolar CO_2 of 33 mmHg. Based on the hemoglobin saturation curve shown below, how much hemoglobin would a comparable person need to have the same amount of oxygen in his/her blood at 14,000 ft above mean sea level. Assume the person at the higher altitude has a partial pressure of alveolar carbon dioxide of 15 mmHg.

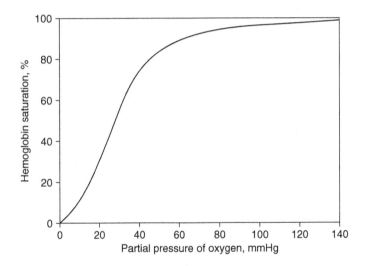

Bibliography

Levitzky MG, *Pulmonary Physiology*, 6th ed., McGraw-Hill, New York; 2003.

Lingappa VR and Farey K. *Physiological Medicine: A Clinical Approach to Basic Medical Physiology*, 1st ed., McGraw-Hill/Appleton & Lange, New York; April 11, 2000.

Martin L. The Four Most Important Equation in Clinical Practice, Case Western Reserve University, http://www.http://www.globalrph.com/martin_4_most2.htm

Richalet JP, Donoso MV, Jimenez D, Antezana AM, Hudson C, Cortes G, Osorio J, and Leon A. Chilean miners commuting from sea level to 4500 m: a prospective study, *High Alt Med Biol*, 2002; 3(2):159–166.

Selkurt EE. *Basic Physiology for the Health Sciences*. Little, Brown and Company, Boston; 1982.

Wark K Jr. and Richards DE, *Thermodynamics*, McGraw-Hill, London; 1999.

West JB. *Respiratory Physiology*, 4th ed, Williams & Wilkins, Baltimore; 1990.

West JB. *Pulmonary Pathophysiology*, 4th ed. Williams & Wilkins, Baltimore; 1992.

4

Hematology and Blood Rheology

4.1 Introduction

Rheology is the study of the deformation and flow of matter. Chapter 4 presents the study of blood, especially the properties associated with the deformation and flow of blood.

4.2 Elements of Blood

Blood consists of 40 to 45 percent formed elements. Those formed elements include red blood cells or erythrocytes, white blood cells or leukocytes, and platelets or thrombocytes. Erythrocytes are those cells involved primarily in the transport of oxygen and carbon dioxide. Leukocytes are cells involved primarily in phagocytosis and immune responses; thrombocytes are involved in blood clotting.

In addition to the formed elements in blood, 55 to 60 percent of blood by volume consists of plasma. Plasma is the transparent, amber-colored liquid in which the cellular components of blood are suspended. It also contains such constituents as proteins, electrolytes, hormones, and nutrients. Serum is blood plasma from which clotting factors have been removed.

4.3 Blood Characteristics

Blood accounts for 6 to 8 percent of body weight in normal, healthy humans. For example, a 165-lb man can expect to have $165 \times (0.07) = 12$ lb of blood. The density of blood is slightly greater than the density of water at approximately 1060 kg/m^3. The increased density comes from the increased density of a red blood cell compared with the density of water or plasma. The density of water is 1000 kg/m^3. Therefore, that 12 lb of blood has a volume of about 12 pt. Most people have between 4.5 and 6.0 L of blood.

In Sec. 1.2.2 we learned about a fluid property known as viscosity. Viscosity is defined by the slope of the curve on a shear stress versus shearing rate diagram. Viscosity of the blood is one of the characteristics of blood that affects the work required to cause the blood to flow through the arteries. The viscosity of blood is in the range of 3 to 6 cP, or 0.003 to 0.006 Ns/m^2. For comparison, the viscosity of water at room temperature is approximately 0.7 cP. Blood is a non-Newtonian fluid, which means that the viscosity of blood is not a constant with respect to the rate of shearing strain. In addition to the rate of shearing strain, the viscosity of blood is also dependent on temperature and on the volume percentage of blood that consists of red blood cells.

The term *hematocrit* is defined as the volume percent of blood that is occupied by erythrocytes, or red blood cells. Since erythrocytes are the cells that make up the oxygen and carbon dioxide carrying capacity of blood, the hematocrit is an important parameter affecting the ability of blood to transport these gases. A normal hematocrit in human males is 42 to 45 percent. Hematocrits below 40 percent are associated with anemia. Those greater than 50 percent are associated with a condition called polycythemia in which the number of red blood cells in an individual is increased above normal.

4.3.1 Types of fluids

A Newtonian fluid is a fluid for which the shear stress and the shearing rate are related by a constant viscosity. Specifically, it means that the viscosity does not vary with the shearing rate. The viscosity of a Newtonian fluid does, of course, vary with temperature as an example. Although blood is not, strictly speaking, a Newtonian fluid, it can be modeled as such in certain circumstances.

$$\text{Newtonian fluid } \tau = \mu\dot{\gamma} \tag{4.1}$$

We see in the following sections that blood behaves as a Newtonian fluid when it flows in tubes that are greater than about 1 mm in diameter and when it flows with rates of shearing strain greater than 100 s^{-1}.

A Bingham fluid, sometimes called a Bingham plastic, is a fluid that has a finite value for shear strength. That is to say, the curve showing stress versus rate of shearing strain for a Bingham fluid does not pass through the origin.

$$\text{Bingham fluid } \tau = \tau_0 + k\dot{\gamma} \tag{4.2}$$

A power law relationship is one in which shear stress depends on the material constant, k, times the rate of shearing strain raised to the power of n. In a power law relationship $\tau = k\dot{\gamma}^n$. Note that viscosity,

μ, is replaced by k in the power law relationship. A Casson fluid is one with some of the characteristics of a Bingham fluid but also one that follows a power law relationship.

$$\text{Casson fluid} \qquad \tau = \underbrace{\tau_o + k^2\dot{\gamma}}_{\text{Bingham fluid}} + \underbrace{2k\sqrt{\tau_o}\sqrt{\dot{\gamma}}}_{\text{power law relation}} \qquad (4.3)$$

Pseudoplastic fluids have no yield stress and are those defined by $\tau = k(\dot{\gamma})^n$. Note that $n = 1$ defines a Newtonian fluid. If $n < 1$, then the fluid is known as a shear-thinning fluid; if $n > 1$, it is known as a shear-thickening fluid.

4.3.2 Viscosity of blood

Viscosity is strongly dependent on temperature, but since humans are maintained at a constant 37°C, the issue of viscosity change with temperature is not significant. On the other hand, blood viscosity varies with such parameters as hematocrit, shear rate, and even vessel diameter. Figure 4.1 shows the shear stress of blood plotted versus the shear rate for blood at a fixed hematocrit. Note that for shear rates above 100 s^{-1}, viscosity becomes constant.

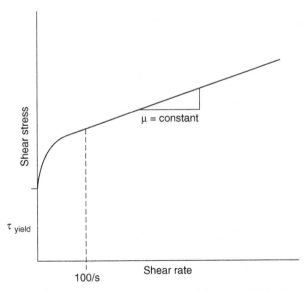

Figure 4.1 Shear stress as a function of the rate of shearing strain (shear rate) for blood.

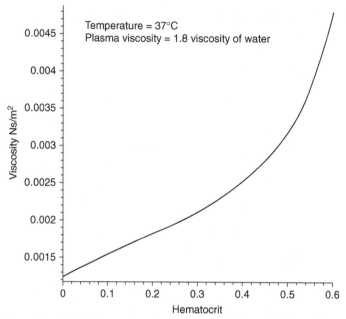

Figure 4.2 Viscosity as a function of hematocrit for whole blood. The graph assumes a temperature of 37°C and plasma viscosity of 1.8 × the viscosity of water at the same temperature. The viscosity of water at 37°C is 0.6915 cP.

The viscosity of blood is also a strong function of hematocrit, or volume percent of red blood cells in the blood. Figure 4.2 shows viscosity versus hematocrit for a viscosity range from 0 to 0.65.

For shear rates greater than 100 s^{-1}, blood behaves as a Newtonian fluid. The yield stress in blood is approximately 0.0003 to 0.03 N/m^2 (Bergel, 1972, pp. 174–175).

4.3.3   Fåhræus-Lindqvist effect

Robert (Robin) Sanno Fåhræus was a Swedish pathologist and hematologist, born on October 15, 1888, in Stockholm. He died on September 18, 1968, in Uppsala, Sweden. Johan Torsten Lindqvist is a Swedish physician, who was born in 1906.

When blood flows through vessels smaller than about 1.5 mm in diameter, the apparent viscosity of the fluid decreases. Fåhræus and Lindqvist published an article in the *American Journal of Physiology* in 1931 describing the effect. This effect is known as the Fåhræus-Lindqvist effect. Figure 4.3 shows the viscosity as a function of diameter and is reproduced from the original 1931 article.

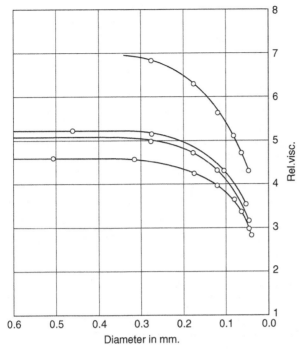

Figure 4.3 Graph from Fåhræus's 1931 paper showing four different trials demonstrating that viscosity of blood depends on the diameter of the tube in which it is flowing. (Used with permission, American Physiological Society.)

As the percent of the diameter of a vessel occupied by the cell-free layer increases, viscosity will decrease (Fig. 4.4). However, when the diameter of the tube approaches the diameter of the erythrocyte, the viscosity increases dramatically.

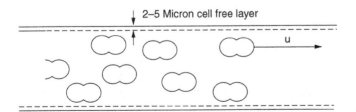

Figure 4.4 Red blood cells flowing through a small-diameter tube. The 2- to 5-μm cell-free layer near the vessel wall contributes to the lowering of viscosity in blood flow through small vessels. U designates the velocity of the individual blood cell.

For blood flow through tubes less than approximately 1 mm in diameter, the viscosity is not constant with respect to tube diameter. Therefore, blood behaves as a non-Newtonian fluid in such blood vessels.

4.3.4 Einstein's equation

Erythrocytes in plasma within a physiological range of hematocrits show a constant shear stress-shear rate relationship typical of a Newtonian fluid. The viscosity of such suspensions can be expressed reasonably well with Einstein's equation for spheres in suspension (Bergel, 1972; Jeffery, 1922; Einstein, 1906).

$$\mu = \mu_p \left(\frac{1}{1 - \alpha\phi} \right) \tag{4.4}$$

where μ is whole blood viscosity, μ_p is plasma viscosity, ϕ is the hematocrit, and α is defined in Eq. (4.5).

$$\alpha = 0.076 \exp\left[2.49\phi + \frac{1107}{T} \exp\left(-1.69\phi \right) \right] \tag{4.5}$$

where ϕ is hematocrit and T is the absolute temperature of the blood in degrees Kelvin.

This equation provides a tool for estimating blood viscosity at various temperatures and hematocrits based on the viscosity of the plasma in which the red blood cells are suspended. The plot in Fig. 4.2, showing viscosity as a function of hematocrit, is generated using this equation.

4.4 Viscosity Measurement

There are several methods of measurement of viscosity. Each of the various methods has its own advantages and disadvantages, which are discussed as the method is presented. The rotating cylinder viscometer is a tool that can be used to measure viscosity and help to strengthen the understanding of viscosity measurement.

4.4.1 Rotating cylinder viscometer

Figure 1.9 in Chap. 1 shows a diagram of a rotating cylinder viscometer. The viscometer consists of a fixed inner cylinder surrounded by a rotating outer cylinder. The gap between the two cylinders contains a fluid whose viscosity can be measured. To measure viscosity, the outer cylinder rotates causing the fluid to move and to generate a torque on the inner cylinder. The torque on the inner "fixed" cylinder causes it to rotate slightly. The rotation of the inner cylinder is opposed by a rod or wire that twists slightly. The inner "fixed" cylinder rotates until the

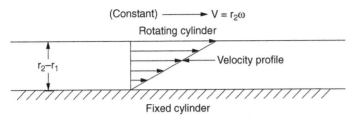

Figure 4.5   Cross section of the rotating cylinder viscometer.

torque created by the shear force in the fluid equals the resisting torque in the wire (Fig. 4.5).

At the point when the outer cylinder is rotating with a constant angular velocity, and the inner cylinder is rotated to some equilibrium position, the wall shear stress in the fluid is equal to the torque in the wire T.

The differential torque exerted on the inner cylinder is given by Eq. (4.6):

$$dT = da \cdot \tau \cdot r_1 = lr d\theta_1 \cdot \tau \cdot r_1 = lr_1^2 \tau d\theta \qquad (4.6)$$

where dT = differential torque, Nm
da = differential area, m^2
τ = wall shear stress, N/m^2
r_1 = outer radius of the inner cylinder, m
l = height of the inner cylinder, m

The shear stress in the fluid can be found by integrating Eq. (4.6) with respect to theta and solving for the shear stress.

$$\tau = \frac{T}{2\pi r_1^2 l} = \text{shear stress in fluid} \qquad (4.7)$$

The velocity of the inner wall of the outer, rotating cylinder is equal to the inner radius r_2, multiplied by the constant angular velocity ω. The velocity of the fluid equals V at the wall and equals zero at the wall of the inner, fixed cylinder.

The velocity of the fluid varies linearly from 0 to V from inner to outer cylinder (see Fig. 4.5). The length of the gap between the cylinders is the difference between the inner radius of the outer cylinder r_2 and the outer radius of the inner cylinder r_1.

The change in fluid velocity with respect to the radial direction, velocity gradient, now becomes:

$$\frac{dV}{dr} = \frac{r_2 \omega}{r_2 - r_1} \qquad (4.8)$$

One restriction applies to Eq. (4.8). The restriction is that $(r_2 - r_1)/r_1$ is very small. To use this equation, the ratio should be less than approximately 0.05.

From its definition, viscosity can now be calculated from the shear stress and the velocity gradient. See Eq. (4.9).

$$\mu = \frac{\tau}{dV/dr} \tag{4.9}$$

4.4.2 Measuring viscosity using Poiseuille's law

Recall from Sec. 1.4.2 that Poiseuille's law relates the flow through a tube to the pressure drop that drives the flow. In fact, that flow is also dependent on viscosity. Poiseuille's law is represented in Eq. (1.17) by:

$$Q = \frac{-\pi R^4}{8\mu} \frac{dP}{dx} \tag{1.17}$$

By solving Eq. (1.17) for viscosity, it is possible to obtain an equation for viscosity as a function of pressure gradient and flow, which suggests a method for obtaining viscosity from a flow and pressure measurement.

$$\mu = \frac{\pi R^4}{8Q} \frac{\Delta P}{\ell_o} \tag{4.10}$$

Figure 4.6 shows the diagram of an apparatus for measuring flow Q through a tube of radius R while simultaneously measuring the pressure drop, ΔP, along a fixed length represented by ℓ_o. The use of this equation assumes a number of restrictions for Poiseuille flow, which are described in Sec. 1.4.2.

Recall that for Poiseuille flow, the velocity profile across the cross section of the vessel is parabolic and can be defined by Eq. (4.11).

$$V = V_{max} \left[1 - \frac{r^2}{R^2} \right] \tag{4.11}$$

where V represents flow velocity, r is the radius variable, and R is the inside diameter of the tube. Recall that, for a parabolic velocity profile, the average velocity across the cross section is half of the maximum velocity. This also leads to a useful equation related to the velocity gradient for Poiseuille flow:

$$\frac{dV}{dr}\bigg|_R = -\frac{V_{max}}{R^2} 2r \bigg|_R = \frac{-2V_{max}}{R} = \frac{-8\overline{V}}{D} \tag{4.12}$$

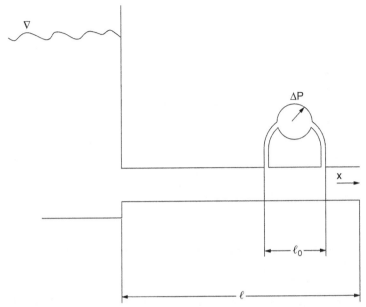

Figure 4.6 An apparatus that can be used to measure viscosity for Poiseuille flow, by measuring flow rate and pressure gradient.

The instantaneous, average velocity across the cross section, $\overline{V}$, is also the same as the flow rate Q divided by the cross-sectional area so that:

$$\frac{dV}{dr}\bigg|_{\text{Wall}} = -\dot{\gamma}_{\text{Wall}} = \frac{-8\,\overline{V}}{D} = \frac{8}{D}\frac{Q}{A} = \frac{4Q}{\pi R^3} \qquad (4.13)$$

This equation also provides a useful tool to estimate the shear rate at the wall in various sized vessels based on their diameter and flow rate or average flow velocity as shown in Table 1.1 in Chap. 1.

4.4.3 Viscosity measurement by a cone and plate viscometer

The cone and plate viscometer works by principles similar to those of the rotating cylinder viscometer, but allows for a blood sample of much smaller volume. A diagram of the cross section of a cone and plate viscometer is shown in Fig. 4.7.

The shear rate is defined by the change in velocity over the gap. For the case of the cone and plate viscometer, the gap δ is a function of r. The velocity of the fluid at the outer wall is also a function of r, so the shear rate can be found from Eq. (4.14).

$$\dot{\gamma} = \frac{\Delta V}{h} = \frac{r\omega}{\delta} = \frac{r\omega}{r\tan\alpha} = \frac{\omega}{\tan\alpha} \qquad (4.14)$$

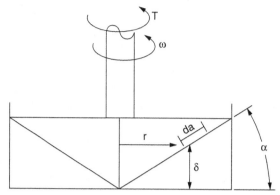

Figure 4.7 Cross-sectional view of a cone and plate viscometer.

If we choose a differential length da along the cone, the circumference around the cone at that point r is $2\pi r$ so that the differential area is equal to $2\pi r da$. The shear stress in the fluid τ multiplied by that differential area gives a force that the fluid generates acting on that tiny area. The shear stress τ is further related to the torque in the shaft T, as shown in Fig. 4.7, in the following way:

$$\underbrace{\underbrace{2\pi r \cdot da}_{\text{area}} \cdot \tau \cdot r}_{\text{force}} = dT \tag{4.15}$$

The variable r is also related to the variable a by the cosine of the angle of the cone (Fig. 4.8). It is therefore convenient to write da in terms of the variable r, as shown in Eq. (4.17).

$$\frac{r}{a} = \cos\alpha \tag{4.16}$$

$$da = \frac{dr}{\cos\alpha} \text{ (puts } da \text{ in terms of } r\text{)} \tag{4.17}$$

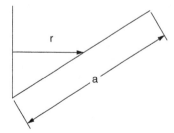

Figure 4.8 The relationship of the variable r, the radius of the cone, and a, some distance along the surface of the cone from the point toward the edge.

Now it is possible to solve for the total torque T by integrating dT over the variable r.

$$T = \int_0^R \frac{2\pi\tau}{\cos\alpha} r^2 dr \tag{4.18}$$

$$T = \frac{2\pi\tau}{\cos\alpha} \frac{R^3}{3} \tag{4.19}$$

Solving Eq. (4.19) for shear stress τ results in Eq. (4.20):

$$\tau = \frac{3}{2} \frac{\cos\alpha}{\pi R^3} T = c_1 T \tag{4.20}$$

The final viscosity is then simply a function of the torque measured, the angle of the cone α, the radius of the cone R, and the speed of rotation ω.

$$\mu = \frac{\tau}{\gamma} = \frac{\dfrac{3}{2}\dfrac{\cos\alpha}{\pi R^3}T}{\dfrac{\omega}{\tan\alpha}} = \frac{3\cos\alpha\,\tan\alpha\,T}{2\pi R^3\omega} \tag{4.21}$$

4.5 Erythrocytes

The word *erythrocyte* comes from the Greek *erythros* for "red" and *kytos* for "hollow," which is commonly translated as "cell." Erythrocytes are biconcave discs with a diameter of approximate 8 μm. Figure 4.9 presents a drawing of a typical erythrocyte showing the basic geometry. Figure 4.10 is a photomicrograph of an erythrocyte. The volume of the typical erythrocyte is approximately 85 to 90 microns^3. The shape of an erythrocyte gives a very large ratio of surface area to volume for the cell. In fact, a sphere with the same volume as a typical red blood cell has only 60 percent as much surface area.

The life span of a red blood cell is approximately 125 days. This means that about 0.8 percent of all red blood cells are destroyed each day, while the same amount of red blood cells are produced in the bone marrow.

Erythrocyte formation is known as erythropoiesis and occurs in the red marrow of many bones including the long bones, such as the humerus and femur, as well as other types of bones, such as the skull and ribs. Iron is required to produce the hemoglobin necessary for red blood cell production and 1 to 4 mg of iron per day is the minimum required. Erythropoeitin stimulates erythrocyte production in response to hypoxia.

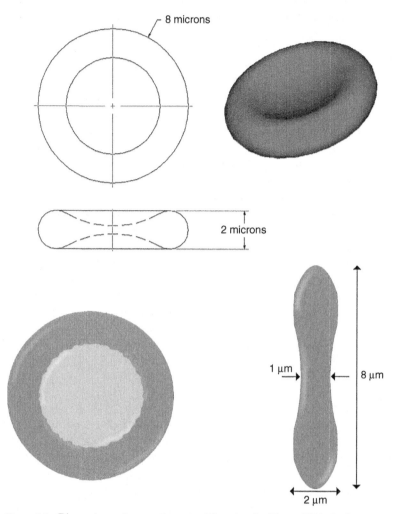

Figure 4.9 Dimensions of an erythrocyte. (*Drawing by Megan Whitaker.*)

Erythrocytes live for approximately 125 days. At the end of the life span of a red blood cell, the erythrocyte becomes fragile and disintegrates. Erythrocytes are destroyed by the macrophages of the mononuclear phagocytic system. The iron is taken to the bone marrow where it is recycled into new hemoglobin.

Normal, healthy human blood typically contains approximately 5 million red blood cells in each cubic millimeter of whole blood. This number depends, of course, on blood hematocrit as well as on the size of the red blood cells. Mature mammalian erythrocytes do not have a nucleus and contain no RNA, no Golgi apparatus, and no mitochondria. Despite that

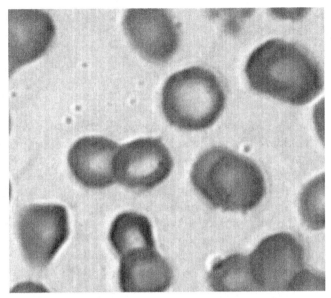

Figure 4.10 Photomicrograph of erythrocytes. Available at http://www.mhprofessional.com/product.php?isbn=0071472177

they do not have mitochondria, erythrocytes are living cells that are metabolically active.

4.5.1 Hemoglobin

Hemoglobin is a large molecule with a molecular weight of 67,000 daltons (Da). A dalton is equivalent to one atomic mass unit, which is equal to one twelfth of the mass of a carbon-12 molecule. Hemoglobin is also a conjugated protein, which means that it is a protein with an attached nonprotein portion. The hemoglobin molecule is composed of four polypeptide subunits (the protein or globin components) and four heme groups (one on each globin chain). At the center of each heme group is an iron atom (Fe^{2+}).

The sigmoid shape of the oxyhemoglobin dissociation curve is a consequence of the four subunits of hemoglobin "cooperating" in the binding of oxygen. At low oxygen pressures, there is a low probability that one or more than one of the four subunits will have an oxygen molecule bound to it. As the pressure increases and oxygen concentration increases, there is an increasing probability that at least one subunit has a bound oxygen. Binding of oxygen to one of the subunits increases the probability that the other subunits will be able to bind an oxygen molecule. So, as oxygen pressure increases even further, the probability that more and more of the remaining binding sites will have an oxygen molecule bound to them rapidly increases.

Hemoglobin readily associates with and disassociates from oxygen and carbon dioxide. Hemoglobin that is associated with oxygen is known as oxyhemoglobin and is bright red in color. Reduced hemoglobin that has been disassociated with oxygen has a bluish purple color. The heme submolecule of the hemoglobin is the portion of hemoglobin that associates with oxygen. Each hemoglobin molecule can transport four O_2 molecules.

The globin submolecule is the protein portion of hemoglobin that associates with carbon dioxide. Each globin submolecule can transport one molecule of CO_2. Normal adult hemoglobin includes three types: type A hemoglobin, type A_2 hemoglobin, and type F hemoglobin. Type A makes up 95 to 98 percent of hemoglobin in adults. Type A_2 makes up 2 to 3 percent, and type F, fetal hemoglobin, makes up a total of 2 to 3 percent. Fetal hemoglobin is the primary hemoglobin produced by the fetus during gestation, and it falls to a much lower level after birth.

S type hemoglobin is associated with sickle cell anemia. Hemoglobin S forms crystals when exposed to low oxygen tension. Sickle cells are fragile and hemolyze, or break down, easily. When the erythrocytes sickle, become fragile, and hemolyze they can block the microcirculation, causing infarcts of various organs. Figure 4.11 shows sickle cell erythrocytes. Persons with S type hemoglobin are also resistant to malaria, which helps this type of hemoglobin to persist in populations where malaria is endemic.

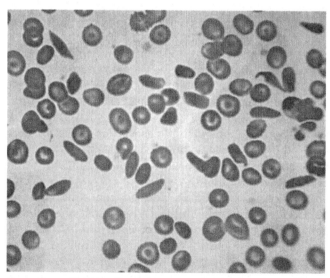

Figure 4.11 Sickle cell erythrocytes. Available at http://www. mhprofessional.com/product.php?isbn=0071472177

4.5.2 Clinical features—sickle cell anemia

Sickle cell anemia presents as severe anemia caused by hemolysis (breaking up of red blood cells). This anemia is punctuated by crises. The symptoms of the anemia are mild in comparison with the severity of the crises because S type hemoglobin gives up oxygen relatively easily compared with hemoglobin A. Some patients have an almost normal life, with few crises. Other patients develop severe crises and may die in early childhood. Sickle cell crises are sometimes very painful. They can be precipitated by such factors as infection, acidosis, dehydration, or deoxygenation (associated with altitude, for example). The most serious vascular occlusive crisis occurs in the brain. Of all patients with sickle cell anemia, 7 percent experience a stroke.

4.5.3 Erythrocyte indices

There can be a number of reasons to cause a person to have a lower than normal quantity of hemoglobin. It could be that the person has a normal number of erythrocytes but lower than normal hemoglobin in each erythrocyte. It could be that the person has a normal amount of hemoglobin in each erythrocyte, but a very low hematocrit. These parameters are quantified and those values are known as erythrocyte indices.

One index measures the average volume of an erythrocyte: mean corpuscular volume (MCV). The MCV can be calculated from the hematocrit of whole blood and the concentration of red blood cells per unit volume as in Eq. (4.22).

$$\text{MCV} = \frac{\text{HCT}}{\text{\# RBC/volume}} \qquad (4.22)$$

Example A patient has a hematocrit of 45 percent red blood cells and has 5 million red blood cells in each cubic millimeter of blood. Estimate the MCV.

$$\text{MCV} = \frac{0.45}{5 \times 10^6 \dfrac{\text{RBC}}{\text{mm}^3} \times \dfrac{1000^3 \text{mm}^3}{\text{m}^3}}$$

$$= 90 \times 10^{-18} \, \text{m}^3 = 90 \, \text{micron}^3$$

$$1 \, \text{micron} = 10^{-6} \, \text{meters}$$

A second index measures the amount of hemoglobin in a red blood cell. That index is known as the mean corpuscular hemoglobin (MCH) index. MCH may be calculated as shown in Eq. (4.23).

$$\text{MCH} = \frac{\text{Hgb/volume}}{\text{\# RBCs/volume}} \qquad (4.23)$$

Example If a patient has a hemoglobin concentration of 15 g/100 mL of whole blood and also has 5 million red blood cells in each cubic millimeter of blood, then the MCH can be calculated in the following manner:

$$\text{MCH} = \frac{15 \text{ g/100 mL}}{5 \times 10^6 \text{ RBC's/mm}^3} = \frac{15 \text{ g/100 mL} \times \dfrac{1 \text{ mL}}{\text{cm}^3} \times \dfrac{\text{cm}^3}{10^3 \text{mm}^3}}{5 \times 10^6 \text{ RBC's/mm}^3}$$

$$= 30 \times 10^{-12} \frac{\text{gm}}{\text{RBC}}$$

A third erythrocyte index is mean corpuscular hemoglobin concentration (MCHC). This index is a measure of the concentration of hemoglobin in whole blood (mass/volume). The MCHC may be calculated as shown in Eq. (4.24):

$$\text{MCHC} = \frac{\text{Hgb/volume}}{\text{Hematocrit}} \tag{4.24}$$

Example For a patient with a hemoglobin concentration of 15 g/100 mL of whole blood and a hematocrit of 0.45, the MCHC can be calculated as shown below:

$$\text{MCHC} = \frac{15 \text{ gm/100 mL}}{.45} = \frac{0.33 \text{ gm of hemoglobin}}{100 \text{ mL blood}}$$

4.5.4 Abnormalities of the blood

Anemia is defined as a reduction below normal in the oxygen-carrying capacity of blood. This reduction in oxygen-carrying capacity could be a result of a decrease in the number of red blood cells in the blood, or it could be a result of a decrease in hemoglobin in each red blood cell, or it could be caused by both.

Macrocytic hyperchromic anemia is an anemia in which a greatly reduced number of red blood cells are too large (macrocytic), contain too much hemoglobin, and are therefore too red (hyperchromic). These large, red erythrocytes lack erythrocyte maturation factor. This lack can sometimes be related to a vitamin B_{12} deficiency.

Polycythemia is defined as an increase in the total red cell mass of the blood. It is commonly thought of as an increase in the number of erythrocytes per cubic millimeter of whole blood above 6 million. Relative polycythemia can be a result of increased concentration of red blood cells due to decreased blood volume, for example, as a result of dehydration. Polycythemia vera, however, results from hyperactivity of the bone marrow. In polycythemia vera there is an abnormal proliferation of all hematopoietic bone marrow elements, often associated with leukocytosis (increase in white blood cells) and thrombocythemia (increase in platelets). It is often difficult to determine the cause of polycythemia vera, but it may be symptomatic of tumors in the bone marrow, kidney, or brain.

4.5.5   Clinical feature—thalassemia

Hemoglobin is a protein carried by erythrocytes. It consists of about 6 percent heme and the rest globin, a colorless protein obtained by removing heme from hemoglobin, the oxygen-carrying compound in red blood cells. A hemoglobin molecule consists of four polypeptide chains: two alpha chains, each with 141 amino acids, and two beta chains, each with 146 amino acids. Thalassemia is a genetic disorder that results from a reduced rate of synthesis of alpha or beta chains in the hemoglobin. In thalassemia the MCV and MCH are low. Traits of alpha thalassemia, failure of genes that produce the alpha chains, are usually not associated with anemia, but the MCV and MCH are low and the red blood cell count is greater than 5.5×10^{12} erythrocytes per liter. The synthesis ratio of alpha:beta chains is reduced below 1:1 in alpha thalassemia.

In beta thalassemia few or no beta chains are produced. Excess alpha chains cause severe ineffective production of red blood cells (erythropoiesis) and hemolysis. The more alpha chains, the more severe the anemia.

Beta thalassemia minor (trait) is common and usually symptomless. MCV and MCH are very low but the red blood cell count is high. Mild or no anemia, with hemoglobin 10 to 15 g/dL, is associated with beta thalassemia minor.

From February 2004 to November 2004, a research project was carried out in Saudi Arabia investigating the occurrence of thalassemia (Al-Suliman, 2006). Healthy subjects coming to six marriage consultation centers in the Al-Hasa area in Saudi Arabia underwent routine mandatory tests. Subjects were considered to have beta-thalassemia trait if they had an MCV < 80 femtoliters (80×10^{-15} L) and/or an MCH < 27 g/dL. The study found that in 3.4 percent of married couples, both partners had the beta-thalassemia trait. The conclusion of the study was that a premarital screening program is helpful in countries with a high incidence of hemoglobin pathologies, for identification and prevention of high-risk marriages. More comprehensive programs were suggested to learn the actual rate of occurrence of beta-thalassemia in that specific region.

4.6   Leukocytes

Leukocytes, also known as white blood cells, can be broadly defined into two groups, arranged by function—phagocytes and immunocytes. They can also be classified into two groups by appearance—granulocytes and agranulocytes. Healthy whole blood normally contains approximately 4000 to 11,000 leukocytes in each cubic millimeter. By comparing that number to 5 million erythrocytes per cubic millimeter, we would expect to see around 500 erythrocytes for every leukocyte.

Leukocytes are translucent. If we look at an unstained blood smear under the microscope, normally we will not see any leukocytes. If we searched very carefully and diligently, perhaps we could find what appears as a white blood cell ghost. Leukocytes each contain a nucleus and other organelles and are easy to find after staining.

The two groups of leukocytes based on function are the phagocytes, which spend their time eating foreign bodies, and the immunocytes, which are involved in the immune response of the body. Phagocytes include neutrophils, eosinophils, monocytes, and basophils. The cells known as immunocytes are lymphocytes.

If we divide white blood cells into groups based on appearance, rather than function, the two groups are granulocytes and agranulocytes. The granulocytes are more granular in appearance compared with the agranulocytes. Granulocytes include neutrophils, eosinophils, and basophils. The agranulocytes include monocytes and lymphocytes. Monocytes are the only phagocytes that are not granulocytes. Table 4.1 includes a listing of types of leukocytes.

4.6.1 Neutrophils

Neutrophils, which are the most abundant type of leukocyte, are granulocytes. Neutrophils have a characteristic appearance with a two- to five-lobed nucleus. Figure 4.12 shows a photograph of a neutrophil. The life span of a neutrophil is around 10 h. Neutrophils are motile and phagocytic, and they play a key role in the body's defense against bacterial invasion. They are the first leukocytes to arrive at an area of tissue damage.

Neutrophil leukocytosis is an increase in the number of circulating neutrophils to a level greater than 7500 per cubic millimeter. Neutrophil

TABLE 4.1   Various Types of Leukocytes Grouped in Order of Their Relative Numbers

Name	Count per cubic millimeter	Size, μm
Neutrophil	2500–7500	10–15
Lymphocyte	1000–3000	10–20
Monocyte	200–800	20–25
Eosinophil	40–400	10–15
Basophil	10–100	10–12
Total leukocytes	4000–11,000	

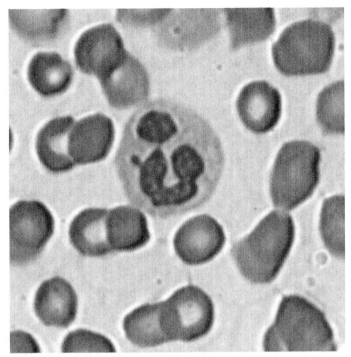

Figure 4.12 Photomicrograph of a neutrophil. (*Note the four-lobed nucleus.*) Available at http://www.mhprofessional.com/product.php? isbn=0071472177

leukocytosis is the most frequently observed change in the blood count. A few causes of this include inflammation and tissue necrosis, for example, as a result of a myocardial infarction, acute hemorrhage, or bacterial infection. Acute infections such as appendicitis, smallpox, or rheumatic fever can result in leukocytosis. If the neutrophil count becomes considerably less than normal, it can also be due to a viral infection such as influenza, hepatitis, or rubella.

Physiologic leukocytosis occurs in newborn infants, pregnancy, and after strenuous exercise, when the number of circulating neutrophils can increase to about 7500 per cubic millimeter with no associated pathology.

4.6.2 Lymphocytes

Lymphocytes are the immunological component of our defense against foreign invasion of the body. Lymphocytes make up about a fourth of the circulating leukocytes. Lymphocytes are nonphagocytic, and a primary function of these cells is the release of antibody molecules and the disposal of antigens.

B lymphocytes originate in the bone marrow and make up about 20 percent of lymphocytes. These cells synthesize antibody molecules. The type of immunity that involves B lymphocytes is known as humoral immunity. The B cells fight against bacteria.

T lymphocytes, on the other hand, originate in the thymus gland and make up about 80 percent of lymphocytes. These T cells are preconditioned to attack antigens either directly or by releasing chemicals to attract neutrophils and B lymphocytes. T lymphocytes do not produce antibodies directly. On the other hand, T cells multiply and clone more T cells, which are also responsive to some antigen. This is known as cell-mediated immunity. T cells fight against bacteria, viruses, protozoa, and fungi, as well as such things as transplanted organs. Patients with acquired immune deficiency syndrome (AIDS) monitor T-cell level, an indicator of the activity of the AIDS virus. Figure 4.13 shows a photomicrograph of a lymphocyte.

Lymphocytosis is defined as an increase in the number of circulating lymphocytes above normal. There are normally 1000 to 3000 lymphocytes in a cubic millimeter of blood. When the number of circulating lymphocytes becomes 10,000 per cubic millimeter, for example, this is known as lymphocytosis. Some causes of acute lymphocytosis include infectious mononucleosis, rubella, mumps, and human immunodeficiency virus (HIV). Causes of chronic lymphocytosis include tuberculosis and syphilis.

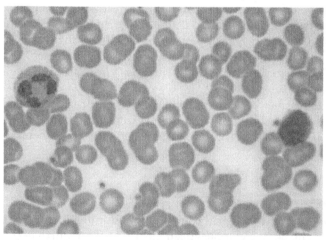

Figure 4.13 Photomicrograph of blood showing a lymphocyte at the right and a neutrophil at the left. Available at http://www. mhprofessional.com/product.php? isbn=0071472177

The life span of a lymphocyte is weeks to months. A large number of inherited or acquired deficiencies in any of the components of the immune system may cause an impaired immune system response. A primary lack of T lymphocytes, as in AIDS, leads not only to bacterial infections, but also to viral, protozoal, and fungal infections.

4.6.3 Monocytes

Monocytes are mobile and actively phagocytic and are larger than other leukocytes. They have a large central oval or indented nucleus. They circulate in the bloodstream for 20 to 40 h and then leave the bloodstream to enter the body's tissues and mature and differentiate into macrophages. Macrophages carry out a function similar to that of a neutrophil. Macrophages may live for several months or even years!

Of all circulating leukocytes, 3 to 9 percent are monocytes. They increase in number in patients with malaria, mononucleosis, typhoid fever, and Rocky Mountain spotted fever.

4.6.4 Eosinophils

Eosinophils are granulocytes that play a special role in allergic response or defense against parasites and in the removal of fibrin formed during inflammation. The primary function of an eosinophil is detoxification of foreign proteins. The number of eosinophils in whole blood is normally 0 to 3 percent of leukocytes. However, in allergic reactions this percentage increases. Eosinophils increase in numbers due to bronchitis, bronchial asthma, or hay fever.

4.6.5 Basophils

Basophils are seen only occasionally in normal peripheral blood.[1] One function of a basophil is to release histamine in an area of tissue damage in order to increase blood flow and to attract other leukocytes to the area of damage. Basophils increase in number in response to hemolytic anemia or chicken pox.

4.6.6 Leukemia

Leukemia is a group of disorders characterized by the accumulation of abnormal white cells in the bone marrow. In other words, there is a purposeless malignant proliferation of leukopoietic tissue, which is the tissue that forms leukocytes. This type of leukopoietic tissue is found in bone

---

[1]Peripheral blood cells are the cellular components of circulating blood, as opposed to the cells kept in the lymphatic system, spleen, liver, bone marrow, or in tissues in case of inflammation.

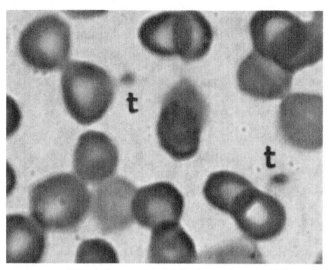

Figure 4.14 Photomicrograph of blood showing thrombocytes, marked t. Available at http://www.mhprofessional.com/product.php? isbn=0071472177

marrow, lymph nodes, spleen, and thymus, for example. Some common features of leukemia include a raised total white cell count, abnormal white blood cells in peripheral blood, and evidence of bone marrow failure.

4.6.7 Thrombocytes

Thrombocytes are also known as platelets. Platelets come from megakarocytes, which are giant (30 μm) cells from bone marrow. The mean diameter of a platelet is about 1 to 2 μm, or about one quarter to one-eighth the diameter of an erythrocyte.

The normal life span of a thrombocyte is 7 to 10 days. The normal platelet count in healthy humans is about 250,000 platelets per cubic millimeter of whole blood. Young platelets spend up to 36 h in the spleen after being released from the bone marrow.

Platelets are granular in appearance and have mitochondria but no nucleus. The main function of platelets is the formation of mechanical plugs during normal hemostatic response to vascular injury. Platelet reactions of adhesion, secretion, aggregation, and fusion are important to this hemostatic function. Thrombocytes are shown in Fig. 4.14 and are marked with a *t*.

4.7 Blood Types

ABO blood types are genetically inherited. To understand blood types, we must first learn a few vocabulary words related to the immune system and genetics. Antibodies, or immunoglobulins, are a structurally related class of glycoproteins that are produced in response to a specific antigen. Antigens, on the other hand, are any substance that attracts antibodies. Antigens include foreign proteins, toxins, bacteria, and viruses.

An allele is any alternative form of a gene that can occupy a particular chromosomal locus. A genotype is the genetic makeup of an individual, or the alleles present at one or more specific loci. A phenotype is the entire physical, biochemical, and physiological makeup of an individual (as opposed to genotype).

The three alleles associated with the ABO blood groups are I^A, I^B, and i. Alleles I^A and I^B are dominant to i, but show no dominance with respect to one another. Each person has two alleles, one inherited from each parent. For example, if a person inherits the I^A allele from each parent, he or she will have the I^A/I^A genotype and blood type A (phenotype).

Agglutination is the term for the aggregation of erythrocytes into clumps. Severe agglutination can lead to death. Agglutinogens (or antigens) are substances on the membranes of erythrocytes. Agglutinins are antibodies in the plasma that try to attack specific agglutinogens. This is known as an antibody–antigen reaction.

Persons with type A blood have the I^A/I^A genotype and their erythrocytes have the A antigen (agglutinogen) and the β or anti-B antibody (agglutinin). If such people receive a transfusion of type B blood, their antibodies would attack the antigens on the donor erythrocytes and cause agglutination.

Persons with type B blood have the B antigen. Persons with type AB blood have both the A and the B antigens on their erythrocytes. Finally, people with type O blood have no A or B antigens on their erythrocytes.

Table 4.2 shows phenotypes and genotypes of the four blood groups along with their respective antigens and antibodies. Note that persons with A blood have the β antibody, persons with type B blood have the α antibody, and persons with O blood have both the α and β antibodies.

TABLE 4.2   Blood Types, Distributions, and Their Associated Antigens and Antibodies

Blood type phenotype	Blood group genotype	Erythrocyte antigen (agglutinogens)	Plasma antibody (agglutinins)	Population distribution whites, Iowa (for example)
A	I^A/I^A or I^A/i	A	Anti-B (β)	42%
B	I^B/I^B or I^B/i	B	Anti-A (α)	9%
AB	I^A/I^B	A and B	Neither	4%
O	i/i	Neither	Anti-A and Anti-B (α & β)	46%

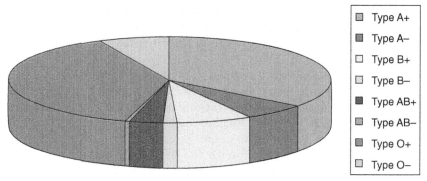

Figure 4.15   Population distribution of ABO blood types among white Iowans.

Figure 4.15 shows the relative population distributions of the various blood types among white North Americans.

People with AB blood have no antibodies and are therefore known as universal recipients. Those with type O blood have both antibodies but no A or B antigens on their red blood cells and are therefore known as universal donors.

H is a weak antigen and nearly all people have it. The I^A and I^B alleles cause H to be converted to the A or B antigen. On the other hand, people with type O blood have more H antigen because it is not converted.

4.7.1   Rh blood groups

Rh antigen was named for the rhesus monkey in which it was first detected. Today, more than 40 Rh antigens are known in humans. One of these antigens is a strong antigen and can lead to transfusion problems.

A person who is Rh negative does not have the Rh antigen and normally does not have the antibody. Upon exposure to the Rh antigen, Rh-negative persons develop the antibody. An Rh-negative mother can produce antibodies after the birth of an Rh-positive neonate after mixing of the mother's blood and the newborn's blood occurs. If that Rh-negative mother with antibodies has a second pregnancy and the fetus is Rh positive, the antibodies will cross over the placenta and begin to destroy fetal erythrocytes (erythroblastosis fetalis). Some of these babies can be saved by transfusion.

Tests can now reveal the presence of the fetal Rh-positive cells in the bloodstream of the Rh-negative mother. When such cells are present, the expectant mother is given injections of the antibody. Because the antibody reacts with and covers up the antigen, the mother's immune system is not stimulated to produce its own antibodies.

TABLE 4.3   MN Phenotypes and Their Associated
Genotypes and Antigens

Phenotype	Genotype	Erythrocyte antigen
M	M/M	M
MN	M/N	M and N
N	N/N	N

4.7.2   M and N blood group system

Many other antigens exist, and although they may not be medically important in transfusions, they can be used to study genetics and are used in tests involving disputed parenthood.

Two antigens, M and N, exist and the alleles for production of these antigens show no dominance to each other. There are therefore three corresponding genotypes: M, MN, and N. The M phenotype is the blood type with the M antigen located on the erythrocyte and is associated with the M/M genotype. Table 4.3 shows a list of MN phenotypes along with their respective genotypes and antigens.

4.8   Plasma

Plasma is a transparent amber fluid and is 90 percent water by volume. The 10 percent that is not water adds some very important characteristics to the function of blood.

Plasma contains inorganic substances like sodium ions, potassium ions, chloride ions, bicarbonate ions, calcium ions, and those chemicals make up about 1 percent of the plasma by volume. Plasma proteins make up about 7 percent of plasma by volume. Those proteins include albumins, globulins, and fibrinogens. In plasma there is a further ~1 percent by volume of nonprotein organics and also varying amounts of hormones, enzymes, vitamins, and dissolved gases.

Plasma proteins are large molecules with high molecular weight that do not pass through the capillary wall. They remain in the blood vessel and establish an osmotic gradient.

Albumin accounts for most of the plasma protein osmotic pressure. Human plasma contains about 4 to 5 g albumin/100 mL plasma. Albumin is important in binding certain substances that are transported in plasma, such as barbiturates and bilirubin.

Globulins are present in human plasma at the rate of approximately 2 to 3 g/100 mL plasma. The normal albumin-to-globulin ratio is approximately 2. Fibrinogen is a plasma protein that is involved in hemostasis, which is the process by which loss of blood from the vascular system is reduced (blood clotting).

4.8.1 Plasma viscosity

Plasma viscosity is a function of the concentration of plasma proteins of large molecular size. This is particularly true of the proteins with pronounced axial asymmetry. Fibrinogen and some of the immunoglobulins fall in this category. Normal values of viscosity at room temperature are in the range of 1.5 to 1.7 cP (1.5 to 1.7×10^{-3} Ns/m^2).

4.8.2 Electrolyte composition of plasma

One equivalent (eq) contains Avogadro's number of + or − charges (6×10^{23} charges). Therefore, for singly charged (univalent ions) 1 equivalent of ions is equal to 1 mole and therefore 1 meq = 1 mmol.

Table 4.4 shows the electrolyte composition of plasma. Note that sodium is an important ion in plasma, which contains 142 meq/L. This is the same as 142 mol/L. Plasma also contains 5 meq/L of calcium ion, which is the same as 2.5 mmol/L because calcium is a doubly charged ion.

Osmotic pressure is defined as the pressure that builds up as a result of the tendency of water to diffuse down the concentration gradient. On the other hand, oncotic pressure is osmotic pressure due to plasma proteins.

Osmolarity, in units of Osm/L, is the measure of the concentration of a solute that would cause an osmotic pressure. Osmolarity is equal to molar concentration multiplied by the number of ionized particles in the solution. When a solute has the concentration of 1 Osmole (Osm), it would have an osmotic pressure of 22.4 atm compared with pure water. Human plasma has the same osmotic pressure as 0.9 percent NaCl solution (physiologic or isotonic saline).

The 0.9 percent NaCl is a solution that contains 9 g of NaCl in 1000 g of water or 9 g NaCl/L water; 1 mol of NaCl contains 23 g of sodium and 35 g of chlorine for a total of 58 g of sodium chloride. The molar concentration of isotonic saline is 9 g of sodium chloride in 1 L of water or 9/58 = 155 mmolar. The osmolarity of isotonic saline is 155 mmolar multiplied by 2 ionized particles (sodium and chloride) or 310 mOsm.

TABLE 4.4 Electrolyte Composition of Plasma

Substance	Symbol	meq/L	Substance	Symbol	meq/L
Sodium	Na$^+$	142	Chloride	Cl$^-$	103
Potassium	K$^+$	4	Bicarbonate	HCO$_3^-$	28
Calcium	Ca^{++}	5	Proteins		17
Magnesium	Mg$^+$	2	Others		5
Total		153			153

When erythrocytes are placed in an isotonic solution, they do not change size. When they are placed in a hypertonic solution, water passes out of the cell and they shrink. When the erythrocytes are placed in a hypotonic solution (0.35 percent NaCl, for example), they swell to a nearly spherical shape and in some cases undergo hemolysis.

4.8.2   Blood pH

The pH of normal healthy blood is in the range of 7.35 to 7.45. When the pH is less than 7.35, this condition is defined as acidosis. When the pH is greater than 7.45, this condition is defined as alkalosis. CO_2 dissolved in water in plasma produces carbonic acid, which lowers blood pH. Bicarbonate and carbonic acid form an acid–base buffer pair, which helps to keep the arterial pH near 7.4. When P_{CO_2} decreases, then the pH increases.

The ratio of bicarbonate concentration to partial pressure of CO_2 in blood is a ratio of metabolic compensation to respiratory compensation. The Henderson–Hasselbalch equation can be used to calculate the pH of arterial blood based on that ratio.

$$pH = 6.1 + \log \frac{[HCO_3^-]}{0.03(P_{CO_2})} \tag{4.25}$$

where the bicarbonate concentration is given in milliequivalents per liter and the partial pressure of carbon dioxide is given in millimeters of mercury. The normal ratio of metabolic compensation to respiratory compensation can be estimated as follows:

$$\text{Normal ratio} = \frac{(HCO_3^-)}{0.03\,(P_{CO_2})} = \frac{24}{0.03(40)} = 20$$

4.8.3   Clinical features—acid–base imbalance

An imbalance of acid generation or removal due to either lungs or kidneys is usually compensated for by a nearly equal and opposite change in the other organ. For example, diarrhea with a loss of bicarbonate-containing fluids will trigger hyperventilation and respiratory compensation to eliminate enough CO_2 to compensate for the bicarbonate loss in the stool. Conversely, vomiting with loss of stomach hydrochloric acid results in metabolic alkalosis for which hypoventilation and an increase in blood CO_2 is the compensation.

Review Problems

1. Whole blood (assume $\mu = 0.004$ Ns/m^2) is placed in a concentric cylinder viscometer. The gap width is 1 mm and the inner cylinder radius is 30 mm. Estimate the torque exerted on the 10-cm-long inner cylinder. Assume the angular velocity of the outer cylinder to be 30 rad/min.

2. Whole blood is forced from a large reservoir through a small rigid tube (diameter = 2 mm; length = 500 mm) and discharges into the atmosphere. If the gauge pressure in the reservoir is 4×10^4 N/m^2, estimate the discharge from the tube in m^3/s. The blood hematocrit is 55 percent and the blood temperature is 30°C. Assume the relative viscosity of plasma to be 1.24×10^{-3} Ns/m^2.

3. For a certain cone and plate viscometer the cone angle is 1.5°. A fluid is placed in the device and the following data are obtained. Is the fluid Newtonian? Explain. Determine the viscosity at the lowest speed in Ns/m^2.

Shear stress dyn/cm^2	Speed of rotation rad/min
19.18	60
9.59	30
3.84	12

4. Blood flowing through the descending aorta can become turbulent in some highly trained athletes. Does blood act as a Newtonian fluid in this case? Why or why not?

5. Using the Einstein equation to predict viscosity as a function of hematocrit and temperature, calculate the viscosity of blood at 37°C using a hematocrit of 42 percent and a plasma viscosity of 1.24×10^{-3} Ns/m^2.

6. Using the Einstein equation to predict viscosity as a function of hematocrit and temperature, calculate the viscosity of blood at 37°C using a hematocrit of 50 and a plasma viscosity of 1.24×10^{-3} Ns/m^2. What is the percent increase in viscosity compared to the value in Prob. 5?

7. In looking in a microscope field at whole blood, how many erythrocytes do you expect to see for each leukocyte? How many for every thrombocyte? How many for every neutrophil?

8. A patient with thalassemia trait has a blood test that shows a hematocrit of 40 percent and a red blood cell concentration of 5.5×10^6 RBC/mm^3. Determine the mean corpuscular volume for this patient.

9. If the same patient in Prob. 8 has a measured hemoglobin of 10 g/100 mL, calculate the mean corpuscular hemoglobin and the mean corpuscular hemoglobin concentration.

10. Calculate the ratio of metabolic compensation to respiratory compensation for a person with a bicarbonate concentration of 24 meq/L and a P_{CO_2} of 25 mmHg. Does this person have acidosis or alkalosis? Respiratory or metabolic?

11. List the components of blood.

12. Describe the Fåhræus–Lindqvist effect.

13. Blood is forced from a large reservoir through a small rigid tube (diameter = 2 mm, length = 500 mm) and discharges into the atmosphere. If the gauge pressure in the reservoir is 2×10^4 N/m^2, estimate the discharge from the tube in m^3/s. Given:

    blood hematocrit = 45 percent

    blood temperature = 37°C

    viscosity of plasma = 0.00125 Ns/m^2

    blood density = 1060 kg/m^3

14. Blood is forced from a large reservoir through a small rigid tube (diameter = 2 mm, length = 500 mm) and discharges into the atmosphere. If the gauge pressure in the reservoir is 2×10^4 N/m^2, estimate the discharge from the tube in m^3/s. Given:

    blood hematocrit = 50 percent

    blood temperature = 37°C

    viscosity of plasma = 0.00126 Ns/m^2

    blood density = 1060 kg/m^3

15. Estimate the rate of shearing strain for a capillary with a diameter of 100 μm and a mean velocity of 40 mm/s.

Bibliography

Bergel, DH, *Cardiovascular Fluid Mechanics*, Vol. 2, Academic Press, London, 1972, Chap. 15, p. 166.

Einstein, A, Eine neue Bestimmung der Molekueldimensionen, *Ann Phys*, 19:289–306, 1906; with a correction in 34:591, 1911.

Fåhræus, R and Lindqvist, R, Viscosity of blood in narrow capillary tubes, *Am J Physiol*, 96:562–568, 1931.

Hoffbrand, AV and Pettit, JE, *Essential Haematology*, 3rd ed., Blackwell Scientific, Oxford, 1998.

Jeffery, GB, The motion of ellipsoidal particles immersed in a viscous fluid, *Proceedings of the Royal Society of London. Series A, Containing Papers of a Mathematical and Physical Character*, Vol. 102, No. 715. (Nov. 1, 1922), pp. 161–179.

Selkurt, EE, *Basic Physiology for the Health Sciences*, Little, Brown and company, Boston, 1982.

Al-Suliman A, Prevalence of beta-thalassemia trait in premarital screening in Al-Hasa, Saudi Arabia, *Ann Saudi Med*, 26(1):14–16, 2006.

Tassiopoulos, S, Deftereos, S, Konstantopoulos, K, Farmakis, D, Tsironi, M, Kyriakidis, M, and Aessopos, A, Does heterozygous β-thalassemia confer a protection against coronary artery disease? *Ann NY Acad Sci*, 1054:467–470, 2005.

Anatomy and Physiology
of Blood Vessels

5.1 Introduction

Arteries are the high-pressure blood vessels that transport blood from the heart, through increasingly smaller arteries, to arterioles, and further to the level of capillaries. Veins conduct the blood from the capillaries back to the heart on the lower-pressure side of the cardiovascular system. The structure of arteries and veins as well as their mechanical properties are discussed in this chapter. At any given time about 13 percent of the total blood volume resides in the arteries, and about 7 percent resides in the capillaries.

Veins, arteries, and capillaries differ in structure because they are specialized to perform their respective perfusion, exchange, and capacitance function. However, the inner layer of all blood vessels is lined with a single layer of endothelial cells.

5.2 General Structure of Arteries

There are three types of arteries which can be classified according to their structure and size. Elastic arteries, which include the aorta, have a relatively greater diameter and a greater number of elastic fibers. Muscular arteries are smaller in diameter than elastic arteries, but larger than arterioles, and they have a relatively larger proportion of muscle compared to connective tissue. Arterioles are the smallest-diameter arteries and have a few layers of smooth muscle tissue and almost no connective tissue.

In general, arteries are composed of three layers: the **tunica intima** or the innermost layer, the **tunica medica** or middle layer, and the

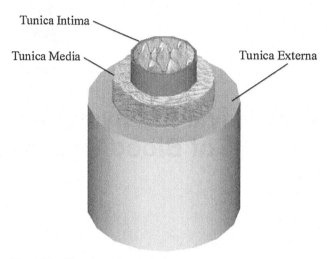

Tunica Intima

Tunica Media

Tunica Externa

Figure 5.1 Structure of an artery.

tunica externa, which is the outermost layer of the artery. The three layers of an artery are shown in Fig. 5.1.

The **lumen** of an artery is the inside of the vessel where blood flows. The lumen is lined with the endothelium which forms the interface between the blood and the vessel wall. The endothelium consists of simple squamous epithelium cells that line the lumen. The cells are in close contact and form a slick layer that prevents interaction of blood cells with the vessel wall as blood moves through the vessel lumen.

5.2.1 Tunica intima

The endothelium plays an important role in the mechanics of blood flow, blood clotting, and leukocyte adhesion. For years, the endothelium was thought of as an inert single layer of cells that passively allowed the passage of water and other small molecules across the vessel wall. Today, it is clear that it performs many other functions, such as the secretion of vasoactive substances and the contraction and relaxation of vascular smooth muscle. The innermost layer of an artery that is composed of endothelium is known as the tunica intima. The tunica intima is one cell layer thick and it is composed of endothelial cells.

5.2.2 Tunica media

The tunica media or middle coat is the middle layer of a blood vessel. The tunica media consists primarily of smooth muscle cells. The tunica media is living and active. The tunica media can contract or expand and change diameters of the vessel, allowing a change in blood flow. Vasoconstriction is the word used to describe a reduction in vessel diameter due to muscular contraction. When smooth muscle in the tunica media relaxes, the diameter of the vessel increases at a given pressure. This increase allows more blood flow for the same driving pressure and this process is known as vasodilation.

The tunica media also consists of elastic tissue. This tissue is passive, does not consist of living cells, and does not have significant metabolic activity. The elastic fibers that make up the tunica media support the blood vessel but also allow recoil of an expanded blood vessel when the pressure is removed. The tunica media is absent in capillaries.

5.2.3 Tunica externa

The tunica externa is also sometimes called the **tunica adventitia.** The tunica externa is composed of connective tissue including passive elastic fibers as in the tunica media. However, the tunica externa also contains passive, much stiffer, collagenous fibers. Just like the tunica media, the tunica externa is absent in capillaries. Figure 5.2 shows a photomicrograph of the cross section of an artery.

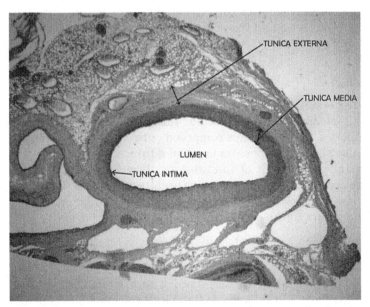

Figure 5.2 Photomicrograph of the cross section of an artery showing the tunica intima, tunica media, and tunica externa. Available at http://www. mhprofessional.com/product.php?isbn=0071472177

5.3 Types of Arteries

Although arteries have the same general structure, they can be divided into groups by their specific functional characteristics. The three types of arteries discussed in this chapter are **elastic arteries, muscular arteries,** and **arterioles.**

5.3.1 Elastic arteries

Elastic arteries have the largest diameter of the three groups of arteries. The aorta is a good example of an elastic artery. One of the chief characteristics of elastic arteries is their ability to stretch and hold additional volume, thus performing the function of a "capacitance" vessel. Just as an electrical capacitor stores charge when given an increased voltage, an elastic vessel can store additional volume when subjected to an increased pressure. Elastic arteries have very thick tunica medias when compared with other arteries and contain a large amount of the elastic fiber elastin.

5.3.2 Muscular arteries

Muscular arteries are intermediate-sized arteries. In these vessels the tunica media is composed almost entirely of smooth muscle. The tunica media of a muscular artery can be as many as 40 cell layers thick. Functionally, muscular arteries can change diameter to influence flow through vasoconstriction and vasodilation. Most arteries are muscular arteries.

5.3.3 Arterioles

Arterioles are defined as those arteries which have a diameter of less than 0.5 mm. Arterioles have a muscular tunica media which is one to five layers thick. These layers are composed entirely of smooth muscle cells. Arterioles also do not possess much of a tunica externa.

Approximately 70 percent of the pressure drop between the heart and the veins occurs in the small arteries and the arterioles. Very little pressure drop occurs in the large arteries, and about 20 percent of the pressure drop occurs in the capillaries.

5.4 Mechanics of Arterial Walls

To understand the mechanics of arterial walls, begin by imagining a long tube of constant cross section and constant wall thickness. This imaginary tube is homogeneous and isotropic (the material properties are identical in all directions). The typical blood vessel is branched and

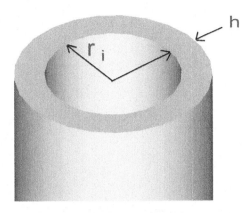

Figure 5.3 An artery modeled as a homogeneous long, straight tube with constant cross section and constant wall thickness.

tapered. It is also nonhomogeneous and nonisotropic. Although our assumptions do not fit the blood vessel, strictly speaking, it makes a practical first estimate of a model to help understand vessel mechanics. We need to continually remember our assumptions and the limitations they bring to the model as we progress in our knowledge of blood vessel mechanics.

Consider a cross section of an artery as shown in Fig. 5.3 with wall thickness h and inside radius r_i. Blood vessels are borderline thin-walled pressure vessels. Thin-walled pressure vessels are those with a thickness-to-radius ratio that is less than or equal to about 0.1. For arteries, the ratio of wall thickness h to inside radius r_i is typically between 0.1 and 0.15, as shown in Eq. (5.1).

$$\frac{h}{r_i} \cong 0.1 \text{ to } 0.15 \tag{5.1}$$

The mechanics of the artery are not dependent only on geometry, but rather to understand the mechanics of the artery we must also consider material properties. In order to consider material properties, let's begin with **Hooke's law** for uniaxial loaded members. Hooke's law for this simple loading condition relates stress to strain in a tensile specimen. Hooke's law for a one-dimensional, uniaxial loaded member is shown in Eq. (5.2).

$$\sigma = E\varepsilon \tag{5.2}$$

where σ is the normal stress in N/m², E is the modulus of elasticity, N/m², and ε is the strain, which is unitless.

Figure 5.4 shows a stress–strain curve for a typical engineering material. The modulus of elasticity is the slope of the stress–strain curve. For

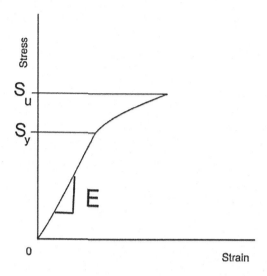

Figure 5.4 Stress–strain curve for a material that is linearly elastic below the yield stress, S_y.

linearly elastic materials, the modulus of elasticity, E, is a constant in the linearly elastic range.

Arteries are not linearly elastic. The stress–strain curve for an artery is shown in Fig. 5.5. It is still possible to define the modulus of elasticity, but the slope of the curve varies with stress and strain. As stress increases in an artery, the material becomes stiffer and resists strain.

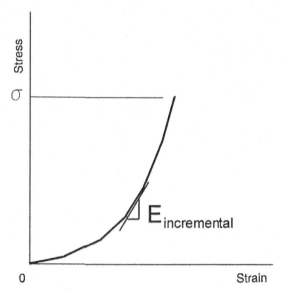

Figure 5.5 Stress–strain curve for an artery.

Arteries are also metabolically active materials. Smooth muscle can contract and expend energy in an effort to resist strain.

Now that we are considering a material in which the modulus of elasticity is not a constant, Hooke's law no longer applies. Instead, the modified Hooke's law as shown in Eq. (5.3) describes the material behavior:

$$\sigma = \int_0^{\varepsilon_f} E_{inc} \, d\varepsilon \tag{5.3}$$

Also, arteries are viscoelastic materials. **Viscoelasticity** is a material property in which the stress is not only dependent on load and area, but also on the rate of strain. For a material in which the stress is dependent on the rate of strain, Eq. (5.4) is true.

$$\sigma = E_1 \varepsilon + E_2 \frac{d\varepsilon}{dt} \tag{5.4}$$

5.5 Compliance

To understand the material property that relates a vessel diameter to the pressure inside that tube, begin by considering a section of the tube shown in Fig. 5.6. In the figure the cross section of a tube is shown, where r_i is the inside radius of the artery, r_o is the outside radius of the artery, and h is the thickness of the tube, $h = r_i - r_o$. Figure 5.7 shows a

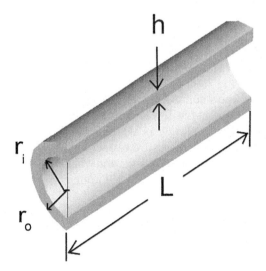

Figure 5.6 Isometric drawing of a tube used to model compliance.

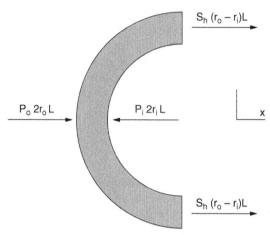

Figure 5.7 Cross-sectional drawing of a tube used to model compliance.

free-body diagram of the same tube on which we will now perform a force balance. Note that S_h is the circumferential hoop stress.

$$\Sigma F_x = 0$$

$$P_o 2r_o L + 2S_h(r_o - r_i)L = P_i 2r_i L \tag{5.5}$$

$$S_h(r_o - r_i) = (P_i r_i) - (P_o r_o) \tag{5.6}$$

$$S_h h = (P_i r_i) - (P_o r_o) \tag{5.7}$$

We now define wall tension T, which is equivalent to the hoop stress in a pressure vessel multiplied by the wall thickness. Wall tension has units of force per length.

$$S_h h \equiv T \text{ wall tension (force per unit length)} \tag{5.8}$$

From mechanics of materials, we know something about the relationship between stress, pressure, and tube geometry. Replacing $S_h h$ with T yields the **law of Laplace** as written in Eq. (5.9). Let us also call $P_i - P_o$ simply transmural pressure P. Now for thin-walled pressure vessels where the outside diameter is approximately equal to the inside diameter, and where the ratio of radius to wall thickness r/h is greater than 10, Eq. (5.10) applies.

$$T = Pr \qquad \text{law of Laplace} \tag{5.9}$$

$$S_h = \frac{Pr}{h} \tag{5.10}$$

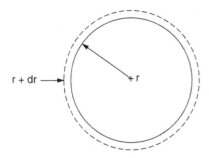

Figure 5.8 Change in vessel radius as a result of increasing pressure.

As the pressure inside the vessel increases, the radius increases, as shown in Fig. 5.8. The resulting incremental hoop strain, $d\varepsilon$, is equal to the change in radius divided by the radius as shown in Eq. (5.11).

$$d\varepsilon = \frac{dr}{r} \tag{5.11}$$

Also, the modulus of elasticity E for thin-walled pressure vessels is equal to the ratio of the incremental hoop stress S_h divided by the incremental hoop strain as shown in Eq. (5.12).

$$E = \frac{dS_h}{d\varepsilon} \tag{5.12}$$

Equations 5.13 and 5.14 relate the initial and final radii and wall thicknesses. The equation is based on conservation of mass. Equation 5.14 assumes constant density for the blood vessel material since blood vessels are essentially incompressible. Equation (5.14) also assumes that the length of the blood vessel does not change significantly.

$$2\pi r_o h_o l_o \rho_o = 2\pi r h l \rho \tag{5.13}$$

$$r_o h_o = r h \tag{5.14}$$

In both equations, r_o is the initial vessel radius and h_o is the initial vessel wall thickness. In both equations, r and h are the vessel radius and wall thickness at any other point in time. Combining Eqs. (5.11) and (5.12), we can now write:

$$dS_h = E \frac{dr}{r} \tag{5.15}$$

By combining Eqs. (5.10) and (5.14), it is also possible to write:

$$dS_h = d\left(\frac{Pr^2}{r_o h_o}\right) \tag{5.16}$$

$$d(Pr^2) = 2Prdr + r^2 dP \tag{5.17}$$

Since for small strains $dr \sim 0$, we can combine Eqs. (5.15), (5.16), and (5.17) into a single equation and solve for the pressure change as a function of radius.

$$E\frac{dr}{r} = \frac{r^2 dP}{r_o h_o} \tag{5.18}$$

$$\int_{P_o}^{P} dP = Eh_o r_o \int_{r_o}^{r} \frac{dr}{r^3} \tag{5.19}$$

$$P - P_o = \frac{Eh_o r_o}{2}\left[\frac{1}{r_o^2} - \frac{1}{r^2}\right] \tag{5.20}$$

$$P - P_o = \frac{Eh_o r_o}{2}\left[1 - \frac{A_o}{A}\right] \tag{5.21}$$

In Eq. (5.21), A_o is the initial cross-sectional area of the vessel and A is the area of the vessel at time t. In Eq. (5.22), we then solve for the area ratio.

$$\frac{A}{A_o} = \left[1 - \frac{(P - P_o)2r_o}{Eh_o}\right]^{-1} \tag{5.22}$$

It is also possible to simplify this expression as a polynomial series since the expression $(P - P_o)2r_o/Eh_o < 1$ for practical, physiological circumstances. The form of the polynomial series is as follows:

$$[1 - x]^{-1} = 1 + x + x^2 + x^3 + \cdots$$

Let $2r_o/Eh_o = C_1$

$$\frac{A}{A_o} = 1 + C_1(P - P_o) + [C_1(P - P_o)]^2 + [C_1(P - P_o)]^3 + \cdots \tag{5.23}$$

Taking the derivative of the cross-sectional area of the vessel with respect to pressure yields Eq. (5.24).

$$\frac{dA}{dP} = A_o C_1 + A_o C_1^2\, 2(P - P_o) + A_o C_1^3\, 3(P - P_o)^2 + \cdots \tag{5.24}$$

Practically speaking, C_1^2, C_1^3, ... are so small that we can now estimate the **compliance** of the vessel by Eq. (5.25).

$$C = \frac{dA}{dP} \approx A_o C_1 = \frac{\pi r_o^3}{E h_o}$$ (5.25)

Finally, it is also possible to estimate modulus of elasticity if the compliance is known. In Eq. (5.26), the modulus of elasticity is derived from Eq. (5.25).

$$E = \frac{2\pi r_o^3}{C h_o}$$ (5.26)

5.5.1 Compliance example

In Fig. 5.9 a tube with modulus of elasticity, E, is anchored between two stationary ports and injected with a known volume of water. For the pressure increase $(P_f - P_o)$, fixed length L_o, tube diameter D, and volume of injected fluid V, calculate the compliance and estimate the modulus of elasticity. The wall thickness of the tube is 1 mm.

Initial pressure, $P_o = 10$ kPa
Pressure increase, $P_f - P_o = 6$ kPa
Initial tube diameter, $D = 20$ mm
Initial length, $L_o = 25$ cm
Volume injected, $V = 10$ mL

The change in cross-sectional area resulting from an injection of 10 mL into the tube is the change in volume divided by the fixed tube length.

$$\Delta A = \frac{\Delta V}{L_o} = \frac{10 \text{ cm}^3}{25 \text{ cm}} = 0.4 \text{ cm}^2$$

The compliance of the tube can then be calculated as the change in area divided by the change in pressure.

$$\text{Compliance} = \frac{dA}{dP} = \frac{0.4 \text{ cm}^2}{6000 \text{ Pa}} \times \frac{1 \text{ m}^2}{(100)^2 \text{ cm}^2} = 6.7 \times 10^{-9} \frac{\text{m}^2}{\text{Pa}}$$

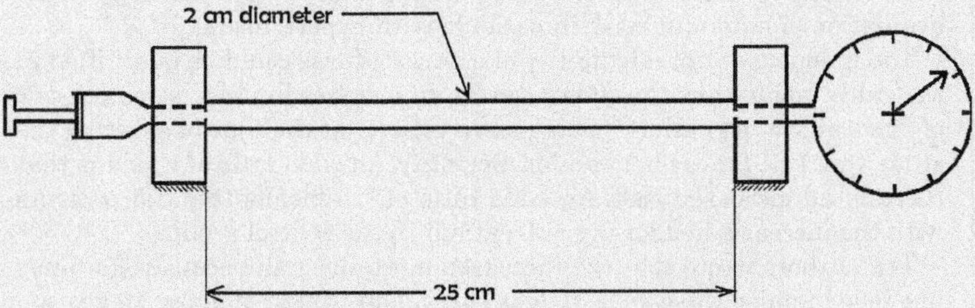

2 cm diameter

25 cm

Figure 5.9 Measuring compliance and estimating modulus of elasticity.

Once the compliance has been determined, the incremental modulus of elasticity for the tube can be estimated from the compliance.

$$\text{Modulus of elasticity} = E = \frac{2\pi r_o^3}{Ch_o} = \frac{2\pi(0.01)^3 \text{m}^3}{6.7 \times 10^{-9}\,\dfrac{\text{m}^2}{\text{Pa}} \times 0.001\,\text{m}} = 0.94\,\text{MPa}$$

5.5.2 Clinical feature—arterial compliance and hypertension

Physicians are interested in vascular stiffness because a number of studies have shown that vascular stiffness is related to hypertension. It is thought, for example, that isolated systolic hypertension, one of the most common types of hypertension among older persons, results from age-associated vascular stiffening and reduced compliance (Schiffrin, 2004).

It is possible to understand and quantify vascular compliance by noninvasive measurements and by analysis of pressure waveforms. From the Moens–Korteweg equation (Sec. 5.6), one can see that pulse wave velocity is directly related to the vessel's modulus of elasticity and hence compliance. Evaluation of pulse wave velocities shows that arterial stiffness increases with advancing age. Arterial stiffness is significantly and independently associated with increased risk for cardiovascular morbidity and mortality.

For example, Laurent et al. (2001) performed a study at Broussais University Hospital in Paris, France, from 1980 to 1996, in which they measured aortic stiffness of a large group of patients regularly over time. When patients entered the study, aortic stiffness was assessed from the measurement of carotid-femoral pulse wave velocity.

Pulse wave velocity was measured along the descending aorta using the foot-to-foot velocity method. Waveforms were obtained transcutaneously over the common carotid artery and the right femoral artery, and the time delay t was measured between the feet of the two waveforms. The distance d covered by the waves was assimilated to the distance measured between the two recording sites. Pulse wave velocity was calculated as the distance traveled by the wave d divided by the transit time t.

The study showed an increased risk of mortality in patients with an increased pulse wave velocity, hence an increased value of modulus of elasticity and increased compliance and that arterial stiffness may help in the evaluation of individual risk in patients with hypertension.

The independent predictability of aortic stiffness could be quantified in the study population: the odds ratio for an increase in pulse wave velocity of 5 m/s above the patient's pulse wave velocity at the time of entering the study was 1.51 for cardiovascular mortality. An odds ratio of 1 means that there is no increased risk. An odds ratio of 1.5 means that 1.5 patients with the increase died for every 1 patient in the control group.

The authors wrote that the increased mortality risk because of a 5 m/s increase in pulse wave velocity was equivalent to that of aging 10 years.

5.6 Pulse Wave Velocity and the Moens–Korteweg Equation

Some years before Otto Frank's attack on the **Windkessel** problem, the relation between wave velocity and elasticity in a tube of fluid was established theoretically by Korteweg in 1878. Dutch mathematician Diederik Korteweg studied at the Polytechnical School at Delft. He attended the University of Amsterdam, receiving a doctorate in 1878. He remained at Amsterdam and became professor there in 1881 until he retired in 1918. For larger arteries the **Moens–Korteweg equation** relates the material properties of the vessel wall to the speed of wave propagation. The simplified version of the equation ignores the modulus of fluid elasticity because it is four orders of magnitude greater than the modulus of elasticity of the vessel wall. The simplified Moens–Korteweg equation becomes:

$$c \cong \sqrt{\frac{Eh}{2r_i\,\rho}}$$

where c is the pulse wave velocity, E is the modulus of elasticity of the vessel wall material, h is the wall thickness of the vessel, r_i is the inside radius of the vessel, and ρ is the density of the blood.

5.6.1 Application box—fabrication of arterial models

Stevanov et al. (2000) reported that they fabricated elastomer arterial models with specific compliance. They made elastic tubes from a two-component silicon elastomer using a lathe for precise control of the wall thickness. The authors measured compliance from quasi-static area–pressure curves and compared those measurements to values of predicted compliance based on the Moens–Korteweg equation. The values of the compliance measurements agreed with the calculated values with an error of 5.5 percent or less.

5.6.2 Pressure–strain modulus

It is difficult to measure the thickness of arteries *in vivo*, so sometimes scientists also report the **pressure–strain modulus**. The pressure strain modulus is defined by the ratio of pressure change to normalized diameter change resulting from that pressure change. Equations (5.27)

TABLE 5.1   Pressure–Strain Modulus of Arteries
in Humans

<35 years old	2.64×10^5 N/m^2
35–60 years old	3.88×10^5 N/m^2
>60 years old	6.28×10^5 N/m^2

through (5.30) designate the pressure–strain modulus, E_p.

$$E_p \equiv \frac{P_{max} - P_{min}}{(d_{max} - d_{min})/d_{mean}} \tag{5.27}$$

$$E = \frac{2\pi r_o^3 2}{h_o \pi d_m}\left(\frac{P_{max} - P_{min}}{d_{max} - d_{min}}\right) \tag{5.28}$$

$$E = 2r_o^2\left(\frac{P_{max} - P_{min}}{d_{max} - d_{min}}\right) \tag{5.29}$$

$$E = \frac{r_o}{h_o} E_p \tag{5.30}$$

In 1972 in their paper titled, "Transcutaneous Measurement of the Elastic Properties of the Human Femoral Artery," Mozersky et al. reported the pressure–strain modulus in humans and how it varies with age. Table 5.1 shows how the stiffness of the blood vessels increases with age.

5.6.3 Example problem—modulus of elasticity

Laurent et al. (2001) performed a study at Broussais University Hospital in Paris, France, from 1980 to 1996, in which they measured the aortic stiffness of a large group of patients regularly over time. When patients entered the study, aortic stiffness was assessed from measurement of carotid-femoral pulse wave velocity.

In the study, the independent predictability of aortic stiffness could be quantified in the study population: the odds ratio for an increase in pulse wave velocity of 5 m/s above the patients' pulse wave velocity at the time of entering the study was 1.51 for cardiovascular mortality.

Problem Estimate the increase in modulus of elasticity of the aorta associated with a 5 m/s increase in the pulse wave velocity. Take the diameter of the aorta to be 2.5 cm and the wall thickness to be 0.25 cm. Use a density of blood of 1060 kg/m^3.

Solution From the Moens–Korteweg equation:

$$c \cong \sqrt{\frac{Eh}{2r_i\rho}}$$

where c is the pulse wave velocity, E is the modulus of elasticity of the vessel wall material, h is the wall thickness of the vessel, r_i is the inside radius of the vessel, and r is the density of the blood.

The predicted increase in modulus of elasticity becomes:

$$\Delta E \cong \frac{2r_i\rho c^2}{h} = \frac{2\frac{2.5}{100}1060(5)^2}{\frac{0.25}{100}} = \boxed{530,000\ \frac{N}{m^2}}$$

5.7   Vascular Pathologies

5.7.1   Atherosclerosis

Atherosclerosis is a vascular pathology that has become a prominent disease in Western society. The term comes from the Greek words *athero* (gruel or paste) and *sclerosis* (hardness), and the disorder is characterized by progressive narrowing and occlusion of blood vessels. When fatty substances, cholesterol, cellular waste products, calcium, and fibrin build up in the inner lining of an artery, this causes a narrowing of the lumen of the vessel and also an increase in the wall stiffness or a decrease in compliance of the vessel. The buildup that results is called plaque. Some level of stiffening of the arteries and narrowing is a normal result of aging. Eventually, the plaque can block an artery and restrict flow through that vessel, resulting in a heart attack if the vessel being blocked is one that supplies blood to the heart. Atherosclerosis can also produce blood clot formation, sometimes resulting in a stroke. When a piece of plaque breaks away from the arterial wall and flows downstream, it can also become lodged in smaller vessels and block flow, also resulting in stroke.

Atherosclerosis usually affects medium-sized or large arteries. Three risk factors associated with atherosclerosis are elevated levels of cholesterol and triglycerides in the blood, high blood pressure, and cigarette smoking.

5.7.2   Stenosis

Stenosis refers to an obstruction of flow through a vessel. When a localized plaque forms inside a vessel, this is called a stenosis. An aortic stenosis, for example, refers to an obstruction at the level of the aortic valve. An aortic stenosis is typically seen as a restricted systolic opening of the valve with an increased pressure drop across the valve. Doppler echocardiography can be used to identify and quantify the severity of a valvular stenosis.

5.7.3 Aneurysm

Aneurysm is the term that refers to the abnormal enlargement or bulging of an artery wall. This condition is the result of a weakness or thinning of the blood vessel wall. Aneurysms can occur in any type of blood vessel, but they usually occur in arteries. Aneurysms commonly occur in the abdomen or the brain. These are known as abdominal aortic aneurysms (AAA) or cerebral aneurysms (CA).

An AAA is an enlargement of the lower part of the aorta. AAAs are known as silent killers because in most cases, there are no symptoms associated with the pathology. When an AAA ruptures, it will result in life-threatening blood loss. Most AAAs are diagnosed during a physical exam when the person is undergoing a checkup for some other health concern.

AAAs that are less than 2 in. in diameter are usually monitored and treated with blood pressure–lowering drugs. AAAs that are greater than 2 in. in diameter are typically surgically replaced with a flexible graft when they are discovered.

A CA is a weak, bulging spot in an artery of the brain. These often result from a weakness in the tunica media (muscle layer) of the vessel that is present from birth. A CA may cause symptoms ranging from headache, drowsiness, neck stiffness, nausea, and vomiting to more severe symptoms such as mental confusion, vertigo (dizziness), and loss of consciousness. A ruptured aneurysm most often results in a severe headache demanding immediate medical attention. A CA can be diagnosed by imaging tests like x-rays, ultrasound, computed axial tomography (CAT), and magnetic resonance imaging (MRI). When detected, brain aneurysms can be treated by microsurgery.

5.7.4 Clinical feature—endovascular aneurysm repair

AAAs are a serious health care issue in the United States with a mortality rate from ruptures of 15,000 patients per year. One common therapy used in conjunction with an AAA is the endovascular stent graft. The endovascular stent graft is a tube composed of fabric supported by a metal mesh called a stent. A minimally invasive technique is used to place the vascular stent graft inside the bulging aneurysm and the stent graft forms the new wall on the aorta, suspended inside of the original aorta. In most cases the stent can be placed using a minimally invasive technique whereby the grafts are inserted through the femoral artery using fluoroscopic control. The graft is released at the appropriate location and expanded with a balloon. The graft is then embedded into the wall of the original aorta using wire barbs.

Some complications of this procedure include endoleaks, graft displacement, and graft migration. Type I leaks occur at the connection between the stent graft and the aorta, with proximal being designated IA and the distal connection being designated IB. Type II leaks occur at a branch of the aorta, outside of the stent graft. Type III leaks are those leaks in which a structural failure of the graft occurs, such as component separation, or those resulting from material fatigue. This type failure can include stent-graft fractures and holes in the fabric of the device.

Some typical endovascular grafts that are or have recently become commercially available include the Medtronic Vascular–AneuRx, which as of 2006 is one of the most commonly implanted. The Gore Exclude from W. L. Gore & Associates is an expanded polytetrafluoroethylene (PTFE) bifurcated graft with an outer self-expanding Nitinol support wire. The Zenith AAA endovascular graft from Cook Vascular, Inc., has woven polyester graft material over stainless steel stents.

The shape of the neck of an aneurysm is an important parameter when considering who is a candidate for an endovascular graft. Hostile neck anatomy is one of the most common reasons that patients are denied endovascular aneurysm repairs (EVAR) (Cox et al., 2006).

Cox et al. report using prophylactic adjunctive proximal balloon-expandable stents in patients with hostile neck anatomy or with type I endoleaks. The neck of an aortic aneurysm describes the point in the aorta at which the vessel begins to balloon. There is a proximal neck and a distal neck. "Adjunctive" describes a secondary therapy which is used in combination with a primary treatment to improve the success of the primary treatment.

Cox's group treated 19 patients with balloon-expandable stents to improve the neck anatomy of the aneurysm. Hostile neck anatomy included a neck length of less than 15 mm and a neck diameter greater than 26 mm. Primary success was achieved in all 19 patients. Based on their experience, the authors recommended that EVAR may be offered to an expanded patient population with hostile neck anatomy with use of prophylactic balloon-expandable stents.

5.7.5 Thrombosis

Thrombosis refers to the formation or development of an aggregation of blood substances, including platelets, fibrin, and cellular elements. Simply put, a thrombus is a blood clot. Thrombosis often results in vascular obstruction at the point of formation. For comparison, an embolism is a general clot or plug that could refer to a blood clot, or an air bubble, or even a mass of cancer cells that is brought by flowing blood from one vessel into a smaller-diameter vessel, where it is eventually lodged.

5.8 Stents

A stent is a metal mesh tube that is inserted into an artery on a balloon catheter and inflated to expand and hold open an artery, so that blood can flow more easily through the artery. The stent remains permanently in place to hold the artery open. Drug-eluting stents are stents that contain drugs that are eluted into the bloodstream and help prevent the arteries from becoming reclogged.

Stents often become necessary when an artery has narrowed due to atherosclerosis. During this vascular disease, plaque builds up on the endothelium. When the vessel becomes 85 to 90 percent blocked, blood flow becomes restricted and the patient may experience symptoms indicating a decrease in blood flow.

Typically, stents are inserted on a balloon catheter through the femoral artery or brachial artery and guided up to the narrowed section of a smaller artery. The balloon catheter allows the stent to be expanded into place, by the inflation of the balloon after the stent has reached the correct location.

5.8.1 Clinical feature—"Stent Wars"

At the time of the writing of this textbook in 2006, drug-eluting stents are the subject of many legal disputes and discussion, which some have referred to as "Stent Wars." Because drug-eluting stents have been so successful in reducing the rate of coronary stent restenosis, marketing analysts have predicted that drug-eluting stents will double the world market for stents to $5 billion annually.

In early 2006 there are two drug-eluting stents which have received Food and Drug Administration (FDA) approval for sale in the United States: the Cordis Cypher™ sirolimus-eluting stent and the Boston-Scientific Taxus™ paclitaxel-eluting stent. The CYPHER received approval in April 2003 and the TAXUS received approval in March 2004.

Sirolimus (rapamycin) is an immunosuppressant drug that is also used to prevent organ rejection after organ transplantation. It is an antibiotic that was first discovered as a by-product of a bacterium found in a soil sample from Easter Island. Rapamycin was originally intended as an antifungal drug, but was abandoned when it was discovered that it had potent immunosuppressive properties in mammals.

Paclitaxel (Taxol) is a drug that has been used in cancer treatment since its discovery in 1967. It was isolated from the bark of the Pacific yew tree, *Taxus brevifolia*. When it was first isolated, paclitaxel was noted for its anti-tumor activity in rodents.

One might ask, "which of the two drug-eluting stents is safer or more effective?" A February 2006 study was done to compare the two stents. It had already been necessary to demonstrate the safety and efficacy of both stents for them to receive their initial FDA approval. However, prior to the 2006 study, there had been no large studies done to evaluate which of the two worked best.

The study (Morice et al., 2006) was conducted in Europe, Latin America, and Asia. It involved 1386 patients at 90 hospitals. Each patient randomly received either the Cypher or the Taxus. Follow-up evaluations were performed after 8 and 12 months.

After a year of stent placement, the rate of adverse cardiac events was similarly low in both groups. The restenosis rates were 9.6 percent for patients who received the Cypher stents and 11.1 percent for the patients who received the Taxus stents. There was not a statistically significant difference in major clinical events including myocardial infarction and target lesion revascularization, a repeat procedure to revascularize the stent location.

With these very good results in both cases, compared to much higher restenosis rates in non-drug-eluting stents, one might think that life is smooth sailing for the two manufacturers. However, there have been difficulties along the way.

In October 2003, the FDA issued a Public Health Notification concerning the Cypher stent. In the notice the FDA wrote:

> Cordis Corporation issued a letter to inform healthcare professionals of a rare but potential risk of thrombosis associated with the use of the CYPHER Sirolimus-Eluting Coronary Stent. The letter provides clarification on the safe use of the product in accordance with the scientific evidence that led to product approval.

The CYPHER stent was approved in April 2003 for patients undergoing angioplasty procedures. Since the product's introduction it is estimated that over 50,000 patients have received a CYPHER stent. To date, FDA has received 47 Medical Device Reports (MDRs) of stent thrombosis occurring at the time of implantation or within a few days of implantation.

In the case of Taxus, in July 2004 the FDA announced a recall of Taxus stent systems as a result of "characteristics in the design of these two lots that resulted in failure of the balloon to deflate and impeded removal of the balloon after stent placement."

5.9   Coronary Artery Bypass Grafting

Coronary arteries are the blood vessels that deliver blood to heart muscle tissue. When atherosclerotic plaque builds up on the wall of the coronary arteries, the result is coronary artery disease. Plaque accumulations can be accelerated by smoking, high blood pressure, elevated cholesterol, and diabetes. In some cases, patients with coronary artery disease can be treated with a minimally invasive procedure called an angioplasty. During an angioplasty, a stent is placed inside the vessel to increase the lumen diameter and to hold the vessel open. When that procedure is not possible, the patient may be a candidate for a coronary artery bypass graft (CABG) surgery. CABG surgery is performed about 350,000 times each year in the United States. When a patient undergoes a bypass graft, the surgeon makes an incision in the middle of the chest and saws through the sternum. The heart is cooled with iced saline while a solution is injected into the arteries to minimize damage.

Veins are typically harvested from the legs to be used as the bypass vessels. It is also possible to use chest arteries, like the internal mammary artery. During the surgery, the bypass vessels are sewed onto the coronary arteries beyond the narrowing or blockage. The other end of the bypass vessel is then attached to the aorta to provide a strong blood supply to the coronary circulation that is unimpeded by blockages.

The entire CABG procedure takes about 4 h to complete. During those 4 h, the aorta is clamped for about 60 min and the blood supply to the body is furnished by cardiopulmonary bypass for about 90 min. At the end of surgery, the sternum is wired together with stainless steel wire and the chest incision is sutured closed.

5.9.1 Arterial grafts

Currently, there are about 700,000 coronary and 70,000 peripheral arterial grafts each year in the United States. Autologous grafts are grafts that are harvested from a patient and grafted to an artery of the same patient. Typical autologous vessels include the greater saphenous vein and the internal mammary artery.

In 1912, Carrel won the Nobel Prize in Medicine for his work on the vascular suture and the transplantation of blood vessels. Then in 1948, Kunlin developed the heart bypass using autologous veins. In 1952, Arthur Voorhees at Columbia University postulated that diseased arteries might be replaced by synthetic fabric. Voorhees and colleagues developed Vinyon-N cloth tubes to substitute for diseased arterial segments. The concept was confirmed and soon synthetic vessels that were larger than 5 mm in diameter were used and often remained patent (unoccluded) for more than 10 years. However, for vessels less than 5 mm in diameter, the prosthetic vessels occlude rapidly upon implantation.

The next advance in vascular grafts came in 1957 when DeBakey and colleagues introduced polyethylene-terephthalate (PET), also known as Dacron®. In 1959, W. S. Edwards introduced the polytetrafluoroethylene (PTFE) graft, also known as Teflon®. Then, 10 years later, in 1969 R. W. Gore invented the expanded polytetrafluoroethylene graft (ePTFE), also known as Gore-Tex®.

The search for a viable, small-diameter vascular graft that is less than 5 mm in diameter continues. By 1978, Herring had described endothelial seeding of synthetic grafts, and by 1987 Cryolife, Inc., began recovering human greater saphenous veins and cryopreserving them for use as vascular grafts. Grafts in which blood vessels come from a human donor and are matched by blood type between the donor and recipient are known as allografts. In 1988 clinical trials of endothelial sodding on ePTFE grafts took place. Sodding is a high-density endothelial seeding in which more than 0.2 million cells per square centimeter are seeded.

One exciting direction for the future of vascular grafts is the area of tissue engineered vessels. L'Heureux et al. (1998) used separately cultured smooth muscle cells and fibroblasts to construct a living blood vessel. The smooth muscle was wrapped around a tube to produce the tunica media. A fibroblast sheet was wrapped around that construct to produce the tunica adventitia. After maturation, the tube was removed and endothelial cells were seeded onto the lumen. Viability was demonstrated.

Review Problems

1. The following values of arterial compliance are reported in the literature:

   Simon et al., 1982 $2.2 \times 10210 \ m^4/N$

   Mergerman et al., 1986   0.05 to 0.20 percent diameter change/mmHg

   Both studies reported values for 5-mm-diameter arteries. Compare these values by converting to m^2/Pa.

2. For the data in Prob. 4, assume a wall thickness of 0.25 mm and estimate the incremental modulus of elasticity.

3. For the data in Prob. 4, assume a wall thickness of 0.25 mm and calculate a pressure–strain modulus for these arteries. Compare these results to E_p as reported by Mozersky et al. (1972). (See Table 5.1)

4. Draw and label the three layers of a blood vessel. Describe the components.

5. The arterial compliance of a vessel is $3.0 \times 10^{-10} \ m^2/Pa$; the artery diameter is 5 mm; the wall thickness is 0.2 mm. Estimate the modulus of elasticity.

6. The modulus of elasticity of a vessel is 2.0 MPa; the artery diameter is 5 mm; the wall thickness is 0.2 mm.

   (a) Estimate the compliance.
   (b) For this vessel, what increase in pressure, in mmHg, would be required to increase the vessel radius by 10 percent?

7. Describe the tunica media. Of what does it typically consist?

8. In the figure below, a tube with modulus of elasticity of 1 MPa is anchored between two stationary ports and injected with a volume of water. The wall thickness of the tube is 1 mm. Estimate the vessel compliance and the volume of water required to raise the pressure by 6 kPa.

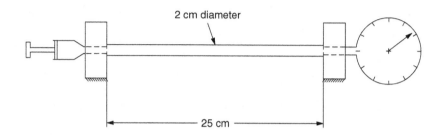

2 cm diameter

25 cm

9. Carotid artery modulus of elasticity is reported by Lin et al. (1999) as 3.12×10^5 N/m^2 and 4.18×10^5 N/m^2 in carotid arterial atherosclerosis, compared to a value of 2.34×10^5 N/m^2 in the control group. Estimate the compliance for both groups.

10. Carotid artery modulus of elasticity is reported by Lin et al. (1999) as 3.12×10^5 N/m^2 and 4.18×10^5 N/m^2 in carotid arterial atherosclerosis, compared to a value of 2.34×10^5 N/m^2 in the control group. Estimate the pulse wave velocity.

11. The diameter of the carotid artery might change from 6.3 to 7.1 mm from diastole to systole. Find the difference in pulse wave velocity predicted by this change in diameter.

Bibliography

Cox, DE, Jacobs, DL, Motaganahalli, RL, Wittgen, CM, Peterson, GJ, Outcomes of endovascular AAA repair in patients with hostile neck anatomy using adjunctive balloon-expandable stents. *Vascular and Endovascular Surgery*. Jan-Feb;40(1):35–40, 2006.

Herrington, DM, Brown, WV, Mosca, L, Davis, W, Eggleston, B, Hundley, G, Raines, J. Relationship between arterial stiffness and subclinical aortic atherosclerosis, *Circulation*, 110, 432–437, 2004.

Laurent, S, Boutouyrie, P, Asmar, R, Gautier, I, Laloux, B, Aortic stiffness is an independent predictor of all-cause and cardiovascular mortality in hypertensive patients, *Hypertension*, 37, 1236–1241, 2001.

L'Heureux N, Paquet S, Labbe R, Germain L, Auger FA. A completely biological tissue-engineered human blood vessel. *FASEB J*. Jan;12(1):47–56, 1998.

Lin, WW, Chen, YT, Hwang, DS, Ting, CT, Want, GK, Lin, CJ, Evaluation of arterial compliance in patients with carotid arterial atherosclerosis, *Zhonghua Yi Xue Za Zhi* (Taipei), 62(9), 598–604, 1999.

Megerman J, Hasson JE, Warnock DF, L'Italien GJ, Abbott WM. Noninvasive measurements of nonlinear arterial elasticity. *Am J Physiol*. Feb;250(2 Pt 2):H181-8, 1986.

Morice, M-C, Colombo, A, Meier, B, Serruys, P, Tamburino, C, Guagliumi, G, Sousa, E, Stoll, H-P, Sirolimus- vs paclitaxel-eluting stents in de novo coronary artery lesions. The Reality trial: a randomized controlled trial, *JAMA*, 295(8), 895–904, 2006.

Moses, JW, Leon, MB, Popma, JJ, et al., SIRIUS Investigators, Sirolimus-eluting stents versus standard stents in patients with stenosis in a native coronary artery, *N Engl J Med*, 349, 1315–1323, 2003.

Mozersky, DJ, Sumner, DS, Hokanson, DE, Strandness, DE, Transcutaneous measurement of the elastic properties of the human femoral artery, *Circulation*, 46, 948–955, 1972.

Schiffrin, E, Vascular stiffening and arterial compliance, implications for systolic blood pressure, *Am J Hypertens,* 17, 39S–48S, 2004.

Selkurt, EE, *Basic Physiology for the Health Sciences*, Little, Brown and company, Boston, 1982.

Simon AC, Levenson JA, Levy BY, Bouthier JE, Peronneau PP, Safar ME, Effect of nitroglycerin on peripheral large arteries in hypertension, Br J Clin Pharmacol. Aug;14(2):241–6, 1982.

Stevanov, M, Baruthio, J, and Eclancher, B, Fabrication of elastomer arterial models with specified compliance. *J Appl Physiol*, Vol. 88, Issue 4, 1291–1294, April 2000.

Stoyioglou, A, Jaff, MR, Department of Cardiology, Lenox Hill Hospital, New York, New York, USA, Medical treatment of peripheral arterial disease: a comprehensive review, *J Vasc Interv Radiol*, 15(11), 1197–1207, 2004.

Voorhees, AB, Jr., Janetzky, A, III, Blakemore, AH, The use of tubes constructed from vinyon 'N' cloth in bridging defects, *Ann Surg,* 135, 332, 1952.

6

Mechanics of Heart Valves

6.1 Introduction

Four cardiac valves help to direct flow through the heart. Heart valves cause blood to flow only in the desired direction. If a heart without heart valves were to contract, it would simply squeeze the blood, causing it to flow both backward and forward (upstream and downstream). Instead, under normal physiological conditions, heart valves act as check valves to prevent blood from flowing in the reverse direction. Also, heart valves remain closed until the pressure behind the valve is large enough to cause blood to move forward.

Each human heart has two **atrioventricular valves,** which are located between the atria and the ventricles. The **tricuspid valve** is the valve between the right atrium and the right ventricle. The **mitral valve** is the valve between the left atrium and the left ventricle. The mitral valve prevents blood from flowing backward into the pulmonary veins and therefore into the lungs, even when the pressure in the left ventricle is very high. The mitral valve is a bicuspid valve having two cusps and the tricuspid valve has three cusps.

The two semilunar valves are the **aortic valve** and the **pulmonic valve.** The aortic valve is located between the aorta and the left ventricle and when it closes it prevents blood from flowing backward from the aorta into the left ventricle. An aortic valve is shown in Fig. 6.1. The pulmonic valve is located between the right ventricle and the pulmonary artery, and when it closes it prevents blood from flowing backward, from the pulmonary artery into the right ventricle.

6.2 Aortic and Pulmonic Valves

The aortic and pulmonic valves consist of three **semilunar cusps** that are embedded within connective tissue. The cusps are attached to a

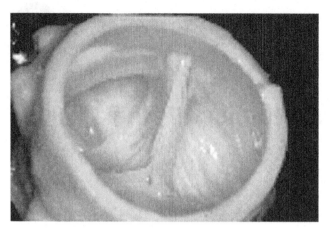

Figure 6.1   Aortic valve. (Reprinted with permission from Lingappa VR, Farey K. *Physiological Medicine*, McGraw-Hill, New York, 2000.

fibrous ring that is embedded in the ventricular septum. The leaflets of the three cusps are lined with endothelial cells and have a dense collagenous core that is adjacent to the aortic side of the leaflets.

A leaflet of an aortic valve is shown in Fig. 6.2. Note the different shape of the leaflets of the mitral valve as shown in Fig. 6.3. The side of the valve leaflet that is adjacent to the aorta is a fibrous layer within the leaflet, and is called the **fibrosa.** The side of the leaflet that is adjacent to the ventricle is composed of collagen and elastin and is called the **ventricularis.** The ventricularis is thinner than the fibrosa and presents a very smooth and slippery surface to blood flow through the valve. The central portion of the valve, known as the spongiosa, contains connective tissue and proteins and is normally not vascularized. The **corpus arantii** (or nodulus of Arantus) is a large collagenous mass in the coaptation region which is thought to aid in valve closure and reduce regurgitation. *Coapt* literally means to approximate, as at the edge of a wound. A coaption surface is one that is formed by two overlapping surfaces that approximate one another.

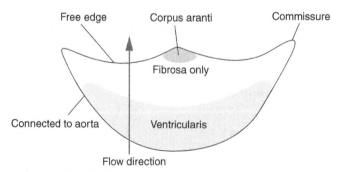

Figure 6.2   Sketch of an aortic valve leaflet.

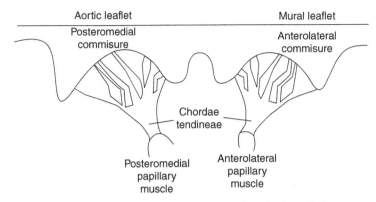

Figure 6.3 Medial view of the dissected mitral valve, cut through the anterolateral commisure.

Collagen fibers in the fibrosa and ventricularis are unorganized in the
unstressed or unloaded condition. When the valve is loaded, however,
the fibers become oriented in a circumferential direction. A fibrous annular ring separates the aorta from the left ventricle.

Three bulges exist at the root of the aorta, immediately superior to
the annular ring of the aortic valve. Those three bulges are known as the
sinus of valsalva or the **aortic sinus.** Two of the sinuses give rise to
the coronary arteries that branch off the aorta. The right coronary artery
comes from the right anterior sinus. The left coronary artery arises
from the left posterior sinus. The third sinus is known as the non coronary sinus or the right posterior sinus.

When the aortic or the pulmonic valve is closed there is a small overlap of tissue from each leaflet that protrudes from the valve and forms
a surface within the aorta. This overlapping surface is once again known
as a coaption surface. This overlapping tissue or coaption surface in the
aortic or pulmonic valve is known as the **lunula.** The lunula is important in ensuring that the valve seals to prevent leakage.

The anatomy of the pulmonic valve is similar to that of the aortic valve.
The chief difference is that the sinuses are smaller and the annulus is
larger in the pulmonic valve. The anatomical dimensions of pulmonic
and aortic valves are reported by Yoganathan et al. in the *Biomedical
Engineering Handbook.* The authors reference a study by Westaby et al.
in 1984 in which 160 pathologic specimens from cadavers were examined.
The mean aortic valve diameter was reported as 23.2 ± 3.3 mm and the
mean pulmonic valve diameter was reported as 24.3 ± 3.0 mm. Another
study that was referenced in the same resource was carried out by
Gramiak and Shah in 1970 and used M-mode echocardiography to measure the aortic root diameter. The aortic root diameter at end systole was

reported as 35 ± 4.2 mm and the aortic root diameter at end diastole was reported as 33.7 ± 4.4 mm.

Diastolic stress in aortic valve leaflets has been estimated at 0.25 MPa for a strain of 15 percent by Thubrikar et al. (1993). The strain in the circumferential direction is about 10 percent of the normal systolic length. There is also an increase in valve surface area associated with diastole. Recall that the aortic valve is relatively unloaded during systole while the valve is open and blood is flowing through it. Larger stresses and strains occur during diastole when the valve is closed and reverse blood flow is prevented.

The aortic valve is a highly dynamic structure. The leaflets open in 20 to 30 ms. Blood then accelerates through the valve and reaches its peak velocity after the leaflets are fully open. Peak blood velocity is reached in the first third of systole and then the blood begins to decelerate after that peak velocity is reached. During the cardiac cycle, the heart also translates and rotates and the base of the aortic valve varies in size, and moves, mainly along the axis of the aorta. The base perimeter of the aortic valve is largest at the end of diastole, at the time during the cardiac cycle when the valve has been closed the maximum length of time. For an aortic pressure range between 120 and 80 mmHg, the perimeter varies approximately 22 percent.

During systole, vortices develop in all three sinuses. These vortices were first described by Leonardo da Vinci (1452–1519) in 1513 and it has been hypothesized that vortices help speed up the closure of the aortic valve. Some of da Vinci's sketches of heart valves are shown in Fig. 6.4. The heart fascinated da Vinci. In his elegant investigation of the valves, he constructed a glass model of the aortic valve and sinuses of valsalva by taking a cast from an ox's heart. Observing the vortices in the sinuses, he concluded that the mechanism of closure of the valves was related to the vortices. When the flow ceased, the vortices pushed against one another to help seal the valve. In fact, it is true that the vortices create a transverse pressure gradient that pushes leaflets toward the center of the aorta.

In healthy individuals, blood flows through the aortic valve at the beginning of systole and rapidly accelerates to a peak value of approximately 1.35 ± 0.3 m/s. This value is slightly higher in children, or about 1.5 ± 0.3 m/s. Pulmonic valve peak velocities are smaller or about 0.75 ± 0.15 m/s in adults and 0.9 ± 0.2 m/s in children (Kilner et al., 1993; Weyman, 1994).

The aortic valve is prone to acquired as well as congenital heart disease. If blood leaks backward through the aortic valve, this condition is known as **aortic regurgitation. A stenosis** is a blockage of blood flow through the valve. Both aortic regurgitation and stenoses are often caused by the calcification of valve tissue.

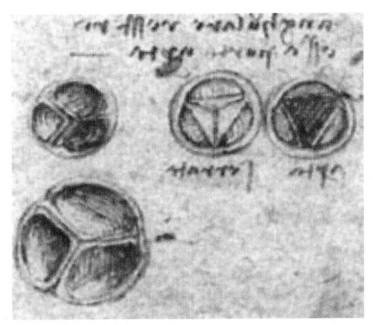

Figure 6.4 Leonardo da Vinci's sketch of an aortic valve. (From Leonardo da Vinci, *The Anatomy of Man*, The Royal Collection © 2006, Her Majesty Queen Elizabeth II.)

6.2.1 Clinical feature—percutaneous aortic valve implantation

Since 1960 when Dr. Albert Starr performed the first long-term successful mitral valve replacement with a caged ball valve, open heart surgery to repair or replace heart valves has become a standard therapy.

Despite the advances that have been made in prosthetic heart valve replacements, some patients are not suitable candidates for surgery. The goal of **percutaneous,** or minimally invasive, valve therapy is to allow for the implantation of heart valves into a patient who is not a candidate for open heart surgery. Percutaneous refers to a medical procedure where access to the heart is done via needle puncture of the skin, rather than by using an "open" approach where the heart is exposed using a scalpel.

The **percutaneous aortic valve** replacement therapy was first developed in 1992. It was initially done with an antegrade procedure, or one in which the direction of insertion of the implant was with the flow. Initially, the valve was delivered on a catheter through the left atrium, into the left ventricle and finally into the aortic root. That initial procedure turned out to be problematic. There were poor outcomes in a number of patients for a number of reasons, particularly in aortic valve replacements. On the other hand there has been significantly better results in pulmonic valve replacement (Feldman, 2006).

At this writing in 2006, the most modern approach to percutaneous aortic valve implantation is a retrograde, or "against the flow" procedure. Webb's group (Webb et al., 2006) describes a new retrograde procedure in which the prosthesis is inserted into the aorta via the femoral artery.

The prosthesis described is a **Cribier-Edwards Percutaneous Valve** from Edwards Lifesciences, Inc. The prosthesis is shown in Fig. 6.5. It is made from a stainless steel stent with an attached trileaflet equine pericardial valve, attached to a fabric cuff. The valve is inserted through the femoral artery to the aorta on a 22F (7.3 mm) or 24F (8 mm) catheter. The catheter is steerable and deflectable allowing manipulation of the prosthesis through the native aortic valve. To reduce cardiac output during deployment of the valve, rapid ventricular pacing was used.

Webb et al. describe the deployment of the percutaneous aortic valve prosthesis in 18 patients with an age of 75 to 87 years for whom surgery was not an option. The implantation was reported to be successful in 14 patients. The aortic valve area increased from an average of 0.6 cm^2 before the procedure to 1.6 cm^2 after the procedure. The authors concluded, "This initial experience suggests that percutaneous transarterial aortic valve implantation is feasible in selected high-risk patients with satisfactory short-term outcomes" (Webb et al., 2006).

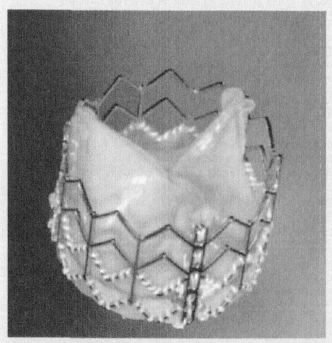

Figure 6.5 Cribier-Edwards Percutaneous Valve. (From Edwards Lifesciences, Inc. AHA Circulation, Lippincott, Williams & Wilkins.)

6.3   Mitral and Tricuspid Valves

Andreas Vesalius (1514–1564) was the Belgian anatomist who was said to have suggested "the picturesque term "mitral" to describe the left atrioventricular valve owing to its resemblance to a plan view of the bishop's mitre" (Ho, 2002). The **mitral** and **tricuspid** valves are composed of four primary elements. These elements include the valve annulus, the valve leaflets, the papillary muscles, and the chordae tendineae. The bases of the leaflets form the annulus which is an elliptical ring of dense collagenous tissue surrounded by muscle. The circumference of the mitral annulus is 8 to 12 cm during ventricular diastole when blood is flowing through the mitral valve from the left atrium to the left ventricle. During ventricular systole the mitral valve closes and the circumference of the mitral annulus is smaller.

Figure 6.6 shows a sketch of a superior view of a mitral valve (the mitral valve as seen from above). The mitral valve is a bicuspid valve and has an aortic (or anterior) leaflet and a mural (or posterior) leaflet. Because of the oblique position of the valve, strictly speaking, neither leaflet is anterior or posterior. The aortic leaflet occupies about a third of the annular circumference and the mural leaflet is long and narrow and lines the remainder of the circumference. When the valve is closed, the view of the valve from the atrium resembles a smile. Each end of the closure line is referred to as a commissure.

Each leaflet is composed of collagen-reinforced endothelium. Striated muscle, nerve fiber, and blood vessels are also present in the mitral valve. Muscle fibers are usually present on the aortic leaflet but rarely on the mural leaflet.

The aortic (anterior) leaflet is slightly larger than the mural (posterior) leaflet. The combined surface of both leaflets is approximately

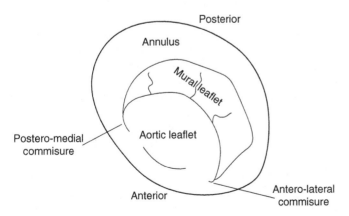

Figure 6.6   Superior view of a mitral valve.

twice the area of the mitral orifice, allowing for a large area of coaptation permitting effective valve sealing.

The mitral valve is shown in a medial view in Fig. 6.3. The valve has been dissected through the anterolateral commissure and laid open. In this view it can be seen that the anterior and posterior leaflets are not separate entities, but rather one continuous piece of tissue.

The chordae tendineae for both leaflets attach to the papillary muscle. Chordae tendineae consist of an inner core of collagen surrounded by loosely meshed elastin and collagen fibers with an outer layer of endothelial cells. There is an average of about 12 primary chordae tendineae attached to each papillary muscle. The anterolateral and the posteromedial papillary muscle attach to the ventricular wall and tether the mitral valve into place. The chordae tendineae prevent the mitral valve from inverting through the annular ring and into the atrium. That is, in normal physiologic situations they prevent mitral valve prolapse. Improper tethering results in prolapse and mitral regurgitation.

Stresses in mitral valve leaflets of up to 0.22 MPa have been estimated when the left ventricular pressure reaches up to 150 mmHg during ventricular systole. Mitral velocity flow curves show a peak in the curve during early filling. This is known as an **E wave.** Peak velocities are typically 50 to 80 cm/s at the mitral annulus. A second velocity peak occurs during atrial systole and is known as an **A wave.**

In healthy individuals, velocities in the A waves are lower than the velocities during the E wave. Diastolic filling of the ventricle shows a peak in flow during diastolic filling followed by a second peak during atrial systole.

6.4   Pressure Gradients across a Stenotic Heart Valve

To assess the seriousness of a heart valve stenosis, it would be helpful to know the area of the stenosed valve when open, or how much of the valve opening is being blocked by the stenosis. Gorlin and Gorlin (1951) developed an equation to empirically predict valve area based on the pressure drop across the valve and the flow rate through the valve. The published equations that were useful for the prediction of area in mitral and aortic valves are still used in cardiology today.

Harris and Robiolio (1999) write, "Although the 'gold standard' for assessment of the severity of mitral stenosis has long been the Gorlin formula derived from data obtained from cardiac catheterization, advances in echocardiography have largely supplanted this procedure. Nonetheless, cardiac catheterization remains a mainstay in the management of mitral stenosis."

6.4.1   The Gorlin equation

We will begin by writing the **Gorlin equation** in terms of variables that are consistent with the other fluid mechanics notations in this textbook. The volume of flow passing through the aortic valve per beat Q is the cardiac output CO divided by the heart rate HR.

$$\text{Flow rate per beat} = \frac{\text{CO}}{\text{HR}} \tag{6.1}$$

The average flow rate through the aortic valve is the same amount divided by the ejection time per beat, T_E.

$$Q = \text{flow rate per beat} = \frac{\text{CO}}{(\text{HR})(T_E)} \tag{6.2}$$

Next, remember that the mean velocity of the flow V (average flow velocity across the flow cross section, through the valve) multiplied by the cross-sectional area A_V is equal to the flow rate.

$$Q = VA_V \tag{6.3}$$

The Bernoulli equation is an equation based on the principle of conservation of energy, in the absence of viscous losses. In fact, there will be significant viscous losses through the heart valve, but the Gorlin equation is an empirical equation intended to account for those losses with an empirically derived valve coefficient. The Bernoulli equation is as follows:

$$P_1 + \frac{1}{2}\rho V_1^2 = P_2 + \frac{1}{2}\rho V_2^2 \tag{6.4}$$

If point 1 referred to in the equation is a reservoir with a very small velocity compared to the jet through the valve, the equation could be rewritten:

$$V_2 = \sqrt{\frac{2\Delta P}{\rho}} \tag{6.5}$$

or, in terms of the volume flow rate:

$$Q = A_V \sqrt{\frac{2\Delta P}{\rho}} \tag{6.6}$$

The above equation relates flow rate, pressure drop, and valve area in the ideal case, but in the case of a real valve, two nonideal characteristics have to be considered. Real blood flow is viscous flow and therefore Eq. (6.5) uses a valve constant, C_V, to represent the actual velocity through the valve.

$$V_2 = C_V \sqrt{\frac{2\Delta P}{\rho}} \tag{6.7}$$

In the second nonideal condition, a jet of blood flowing through the valve will contract to a smaller area than the total cross-sectional valve area. The area of contraction A depends on the valve design and can be designated here using the **valve coefficient** C_C.

$$A = C_C A_V \tag{6.8}$$

Combining Eqs. (6.3), (6.7), and (6.8), the Gorlin equation becomes:

$$Q = C_C C_V A_V \sqrt{\frac{2\Delta P}{\rho}} \tag{6.9}$$

The coefficients are measured for various orifices and it should be carefully noted that both the valve coefficient and the contraction coefficient are empirically derived coefficients which may not be constant under varying pressure and flow conditions as well as varying Reynolds numbers.

In fact, this equation or a modified form of the equation is often used clinically to estimate the effective area of a stenotic heart valve. To further simply its use clinically, the equation is often written in the following form:

$$A_V = \frac{CO}{(T_E \, \text{HR} \, K \, \sqrt{\Delta P})} \tag{6.10}$$

where A_V is the area of the stenotic valve, CO is the cardiac output in cc/min, T_E is the ejection time or the time per beat during which blood flows through the valve. HR is the heart rate in beats per minute, ΔP is the mean pressure gradient over the ejection period in mmHg, and K is the Gorlin constant.

The Gorlin constant is proportional to C_C, C_V, and $1/\sqrt{\rho}$ and contains all the conversion factors necessary to account for a mathematically consistent set of units. The flow rate is equal to:

$$Q = \frac{CO}{T_E \, \text{HR}} \tag{6.11}$$

If the cardiac output is given in cubic centimeters per minute, the ejection period in seconds per beat, and the heart rate in beats per minute, then

Eq. (6.11) yields a flow rate in cubic centimeters per second. It is important once again to note that the Gorlin equation is a useful concept, but that the **Gorlin "constant"** is not a constant but depends on the flow rate.

Therefore, the Gorlin value of the Gorlin constant K is sometimes given as 44.3 in aortic valves and 37.7 in mitral valves. The units associated with the Gorlin constant are as follows:

$$\frac{cm}{s\sqrt{mmHg}}$$

6.4.2 Example problem—Gorlin equation

A patient has a cardiac output of 5 L/min. The systolic ejection period is 358 ms with a heart rate of 70 beats per minute. The mean aortic gradient as measured by echocardiography is 81 mmHg. Find the aortic valve area as estimated by the Gorlin equation. What is the average flow rate across the aortic valve during ejection?

Answer

$$A_V = \frac{CO}{T_E\, HR\, (44.3)\sqrt{\Delta P}}$$

where A_V = aortic valve area, cm^2
   CO = cardiac output, mL/min
   T_E = time of ejection per beat, s
   HR = heart rate, beats/min

mean pressure gradient = left ventricular pressure − aortic pressure, mmHg

$$A_V = \frac{5{,}000}{(0.358)(70)(44.3)\sqrt{81}} = 0.50\ cm^2$$

The average flow rate across the valve is 5000 mL/25 s = 200 mL/s.

6.4.3 Energy loss across a stenotic valve

Energy loss across a stenotic valve is influenced by more than the effective valve area predicted by the Gorlin equation. Garcia et al. (2000) suggest an approach that uses an energy method to more accurately estimate the severity of an aortic stenosis. The authors suggest an index that they refer to as the **energy loss index.** Although it is a measure of energy loss, the energy loss index yields a value with pressure units.

The Gorlin equation uses **transvalvular pressure gradients** to predict **effective orifice areas.** These parameters are commonly used clinically for the assessment of aortic stenoses. Some studies have emphasized the importance of taking into account the pressure recovery phenomenon that occurs downstream from the valve. In other words, the kinetic energy that is stored in the fluid before, and as, it "jets" through the **vena contracta** is converted to pressure as the fluid is decelerated.

The measurement of a **pressure gradient** using Doppler echocardiography is an indirect measurement based on the velocity gradient. Those indirect pressure gradients can differ significantly from the actual pressure gradients measured via a catheter. This phenomenon is often referred to as **"pressure recovery"** and can be significant in patients with moderate to severe aortic stenoses and with small aortic cross-sectional areas.

From the Bernoulli equation, a term the authors describe as **energy loss** E_L can be written by:

$$E_L = P_{VC} - P_A + \frac{1}{2}\rho V_{VC}^2 - \frac{1}{2}\rho V_A^2 \qquad (6.12)$$

where P_{VC} = pressure at the vena contracta
$\quad P_A$ = pressure in the aorta
$\quad V_{VC}$ = velocity of blood in the vena contracta
$\quad V_A$ = velocity in the aorta

Considering linear momentum along the flow direction as shown in Fig. 6.7, the momentum equation can be written as:

$$P_{VC}A_A - P_A A_A = -\rho V_{VC}^2 A_{VC} + \rho V_A^2 A_A \qquad (6.13)$$

Combining Eq. (6.12), the Bernoulli equation between the vena contracta and the aorta, and Eq. (6.13), linear momentum between the vena

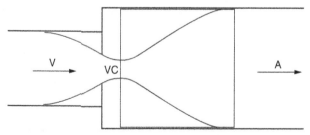

Figure 6.7 Schematic representation of flow through an aortic valve where V is the left ventricle, VC represents the vena contracta, and A represents the aorta. (*Modified from Garcia et al., 2000.*)

contracta and the aorta, to eliminate pressures in the equation yields:

$$E_L = -\left(\rho V_{VC}^2 \frac{A_{VC}}{A_A}\right) + \left(\frac{\rho}{2}V_A^2\right) + \left(\frac{\rho}{2}V_{VC}^2\right) \qquad (6.14)$$

The continuity equation yields:

$$A_{VC} = \frac{A_V V_V}{V_{VC}} \text{ and alternatively } V_A^2 = \frac{V_{VC}^2 A_{VC}^2}{A_A^2} \qquad (6.15)$$

Because measuring the area at the vena contracta is problematic, and because it will be convenient to use $V_V A_V$ multiplied together as cardiac output, the energy loss equation can be rewritten as:

$$E_L = \frac{\rho}{2}\left(V_{VC} - V_V\frac{A_V}{A_A}\right)^2 \qquad (6.16)$$

Now the energy loss can be calculated in terms of variables that are readily measurable through transthoracic echocardiography, as shown in Eq. (6.17).

$$E_L = \frac{\rho}{2}\left(V_{VC} - \frac{CO}{A_A}\right)^2 \qquad (6.17)$$

where ρ = density
V_{VC} = blood velocity at the vena contracta
CO = cardiac output
A_A = cross-sectional area of the aorta
E_L = energy loss expressed in units of pressure

The unitless energy loss coefficient associated with this energy loss is defined by the term:

$$\text{Energy loss coefficient} = \frac{(EOA)(A_A)}{(A_A - EOA)} \qquad (6.18)$$

An energy loss index that is less that $0.5 \text{ cm}^2/\text{m}^2$ should probably be considered a critical value. A value less than 0.5 indicates a serious stenosis.

Recall that the **effective orifice area EOA** is the area at the vena contracta which is derived from the continuity equation, and is shown in Eq. (6.19).

$$EOA = \frac{V_V A_V}{V_{VC}} \qquad (6.19)$$

6.4.4 Example problem—energy loss method

The same patient from the previous example problem using the Gorlin equation has a cardiac output of 5 L/min. Aortic area = 4.9 cm². Density = 1060 kg/m³. Velocity at the vena contracta = 1.66 m/s. The mean aortic gradient as measured by echocardiography is 81 mmHg.

Find the "energy loss" in mmHg and the energy loss coefficient.

$$E_L = \frac{\rho}{2}\left(V_{VC} - \frac{CO}{A_A}\right)^2 = 1060\,\frac{kg}{m^3}\,\frac{N\,s^2}{kg\,m}\left(1.66\,\frac{m}{s} - \frac{5000/60\,\frac{cm^3}{s}}{A_A\,cm^2}\,\frac{m}{100\,cm}\right)^2$$

$$\times\frac{mmHg}{133.3\,\frac{N}{m^2}} = E_L = 13\text{ mmHg}$$

From Eq. (6.19) the effective orifice area is

$$EOA = \frac{V_V A_V}{V_{VC}} = \frac{CO}{V_{VC}} = \frac{5000/60\,\frac{cm^3}{s}}{1.66/100\text{ cm/s}} = 0.50\text{ cm}^2$$

Finally, from Eq. (6.18) the energy loss coefficient is:

$$\text{Energy loss coefficient} = \frac{(EOA)(A_A)}{(A_A - EOA)} = \frac{(0.5)(4.9)}{(4.9 - 0.5)} = 0.46$$

6.4.5 Clinical features

(Lingappa and Farey, 2000)

Chordae tendineae rupture and **papillary muscle paralysis** can be consequences of a heart attack. This can lead to bulging of the valve, excessive backward leakage into the atria (regurgitation), and even valve prolapse. Valve prolapse is the condition under which the valve inverts backward into the atrium. Because of these valve problems, the ventricle does not fill efficiently. Significant further damage, and even death, can occur within the first 24 hours after a heart attack because of this problem.

6.5 Prosthetic Mechanical Valves

Early mechanical prosthetic heart valves were ball valves. In 1952, Dr. Charles Hufnagel implanted the first **ball and cage valve** into a patient with aortic valve insufficiency. Although the design of mechanical prosthetic valves has progressed significantly over the past 50 years, ball and cage valves are still the valve of choice by some surgeons.

Björk-Shiley designed the first tilting disc, or single leaflet, valve around 1969. Those valves used a tilting disc occluder that was held in place by a retaining strut. However, the development of the bileaflet valve by St. Jude Medical in 1978 has led to the predominance of this type of bileaflet valve in the mechanical valve market.

One example of a recently approved mechanical **bileaflet aortic valve** is the On-X valve (Fig. 6.8) manufactured by Medical Carbon Research Institute in Austin, Texas. The On-X was approved by the Food and Drug Administration (FDA) for implant in humans in May 2001. The mitral valve version of the On-X was approved for use by the FDA in March 2002.

The On-X consists of a pyrolytic carbon cage, a Dacron sewing ring fixed by titanium rings, and a bileaflet pyrolytic carbon disc.

Pyrolytic carbon is a material used in artificial heart valves because of its mechanical durability and its characteristic ability not to generate blood clots. The material can be said to have low thrombogenicity. Thrombogenicity is the tendency of a material in contact with the blood to produce a blood clot. Dr. Jack Bokros and Dr. Vincent Gott discovered pyrolytic carbon in 1966. At that time, pyrolytic carbon was used to coat nuclear fuel particles for gas-cooled nuclear reactors. General Atomics initiated a development project headed by Dr. Bokros to develop a biomedical

Figure 6.8   On-X prosthetic made by MCRI, Austin, Texas (a wholly owned subsidiary of Medical Carbon Research Institute Deutschland GmBH, Hanover, Germany; used with permission).

grade of pyrolytic carbon. The result was Pyrolite, which was incorporated into heart valve designs and remains the most widely used material for heart valves today.

6.5.1 Clinical feature—performance of the On-X valve

Between December 2000 and January 2003, Özyurda et al. conducted a study in Turkey to assess the performance of the On-X valve (Medical Carbon Research Institute, Austin, TX). The authors wrote, "Few data are available on the clinical outcome of patients undergoing valve replacement with the On-X valve. Therefore, the aim of this study was to evaluate the early and midterm performance of this new generation bileaflet prosthesis."

On-X valves were implanted into 400 patients. Of these 400, 120 had an aortic valve replacement (AVR), 258 had a mitral valve replacement (MVR), and 22 had a combined aortic and mitral valve replacement (DVR).

Patients were followed up prospectively at 3- to 6-month intervals. The overall hospital mortality in the study was 3.5 percent. When examining the patients 4 years after the initial surgery, the overall survival rate was 92 ± 1 percent. Freedom from thromboembolism and thrombosis events at 4 years in the study were as follows:

- Thromboembolism
  - 99.1 percent for aortic valve replacement
  - 98.3 percent for mitral valve replacement
  - 94.7 percent for dual valve replacement patients
- Thrombosis
  - 100 percent for aortic valve replacements
  - 99.2 percent for mitral valve replacements
  - 94.7 percent for dual valve replacements

The authors also wanted to establish data concerning pressure drops across the On-X. They performed an echocardiograph within 1 year of the aortic valve replacement and found that the mean aortic valve gradient was 12.8 ± 6, 10.3 ± 3, 9.0 ± 4, 8.3 ± 3, and 6.2 ± 3 mmHg for 19, 21, 23, 25, 27/29 mm valve sizes, respectively. The mitral valve replacement mean gradient was 4.9 ± 2, 4.5 ± 1.2, and 4.0 ± 0.8 mmHg for 25, 27/29, 31/33 mm valve sizes, respectively.

The conclusion of the authors after this study was that the "On-X valve is a highly effective mechanical valve substitute with low morbidity and mortality and good functional results."

6.5.2 Case study—the Björk-Shiley convexo-concave heart valve

The design and development of the Björk-Shiley convexo-concave (BSCC) heart valve is interesting from a biomedical engineering design as well as from an ethical standpoint. In the early 1960s the earliest human prosthetic heart valves were developed. In 1976, Dr. Viking Björk and Shiley Incorporated developed the 60° convexo-concave BSCC-60 heart valve. The 60° designation indicates that the valve was designed to have a maximum 60° opening. The valve had improved flow characteristics over other valves that were on the market at the time. In 1979, Shiley came out with a valve that opened even farther. It was the BSCC-70 heart valve. The valve had a cobalt chrome alloy housing with a pyrolytic carbon disc, a Teflon sewing ring, and a cobalt chrome retaining strut which retained the disc in the valve.

During clinical trials of the valve, the first valve failure occurred when the outlet strut fractured. Shiley Inc. reported that failure as an anomaly and the FDA approved the valve for use in the United States. After the BSCC-60 went on the market, outlet strut fractures began to occur. Most of the reported fractures were fatal and between 1980 and 1983 Shiley implemented three voluntary recalls of the valve. During that time, Shiley's share of the prosthetic valve market was up to 60 percent.

In the early 1980s the FDA began an investigation of the Björk-Shiley valve. Evidence later surfaced that Shiley Inc. attempted to delay and conceal damaging information about its product. For example, the FDA requested sets of randomly selected, recalled valves for testing. Instead of sending randomly selected valves, Shiley handpicked the valves that were sent to the FDA for testing.

- Case report 1: 53-year-old German woman in 1992 in Göttingen. See Chap. 2, Sec. 2.2 for a complete description of this case report.

- Case report 2: 49-year-old Japanese businessman in San Francisco in 1990. See below.

Clinical feature—case report. In 1990, a 49-year-old Japanese businessman who was attending a conference in San Francisco was brought to the emergency room of a local hospital (Chen et al., 1992). The patient was suffering from sudden onset severe difficulty breathing with anterior (front) chest pain radiating to his back. The doctors noted that he had a healed, midline sternal scar indicating previous chest surgery. The patient lacked awareness of his surroundings and he was blue around the lips. When the physicians listened to his heart and lung sounds, he had diffuse rales over both lungs and an S_3 gallop. There were no murmurs. High doses of dopamine were required to raise his systolic blood pressure to 70 mmHg. An electrocardiogram showed atrial fibrillation with a 130 beats per minute ventricular rate.

An emergency echocardiogram showed vigorous left ventricular contraction and a bulging atrium. It also showed a prosthetic heart valve but no evidence of prosthetic disc motion. It was subsequently learned that the patient, years earlier, had been implanted with a 60°-opening BSCC prosthetic heart valve.

The man was immediately transferred to St. Mary's Medical Center for emergency cardiopulmonary bypass. A chest x-ray showed the prosthetic valve disc in the patient's descending thoracic aorta. The valve strut was later found in the left ventricle. The damaged prosthesis was explanted and replaced with a 29-mm St. Jude Medical mitral valve.

Three days later the man underwent another surgery for the removal of the prosthetic disc. He recovered fully without complications and was discharged from the hospital 2 weeks later.

In the end, 89,000 people worldwide received BSCC valves. By November 1991, 466 outlet strut fractures had been reported. A large follow-up study of 2,303 60° or 70° convexo-concave valves implanted in the Netherlands was published in 1992 (van der Graaf et al., 1992). It was determined that most patients who died with a BSCC valve had died outside the hospital. Only 24 percent of the patients who had died with the BSCC valves implanted had been examined for strut failure. In the study it was determined that the cumulative failure rate for BSCC60 valves was 4.2 percent over 8 years. The cumulative failure rate for BSCC70 valves was 17.4 percent over 8 years. When a strut fracture occurs in a BSCC heart valve, the mortality rate is very high. For aortic valve with strut failure, 85 percent of the patients die. The overall mortality rate for strut fractures is 75 percent. Shiley Inc. voluntarily withdrew the BSCC heart valve from the market in 1986.

In a January 1994 study published in *Lancet* (de Mol et al., 1994), 24 valves were replaced in 22 patients who were considered to be at high risk with respect to failure of a BSCC heart valve. Of the 24 valves, 7 had experienced a single leg fracture of the strut. The authors wrote, "This finding supports our hypothesis that, owing to the lethal character of the failing aortic-valve strut, fractures remain under-reported, few such patients reach hospital and necropsy is rarely done."

In a January 2000 report (Steyerberg et al., 2000), it was estimated that 35,000 patients were still alive with one or more implanted BSCC valves. Pfizer purchased Shiley in 1979 at the onset of its convexo-concave valve ordeal. In 1992, after years of litigation, Pfizer sold Shiley's businesses to Italy's Sorin Biomedical. Sorin opted not to purchase rights to the convexo-concave valve. Later, a firm by the name of Alliance Medical Technologies, comprising former Shiley and Pfizer employees, purchased the rights to Shiley's Monostrut heart valve line from Sorin. The Monostrut is still available outside the United States. It has developed a good reputation for reliability and good hemodynamics. There are no welded struts on the Monostrut valve as both the inlet and outlet struts are part of the same piece of metal as the valve ring.

Patients with an implanted BSCC heart valve seem to be limited to only a few options:

1. They can elect to keep the valve and hope it does not fail. So far the published failure rates are in the 2 to 17 percent range. The risk of failure is thought to be dependent on fatigue and therefore cumulative with time. In the long run, the failure rates may be much higher.
2. Patients may elect preventive valve replacement surgery. The risk of death for a 50-year-old man from valve replacement surgery is about 5 percent. The cost of such a replacement surgery is conservatively in excess of $25,000 in the United States.

In January 1988, the following letter from the FDA was sent to physicians to provide information about the risk of outlet strut fractures.

January 1998
Dear Doctor:
This letter provides new information about the risk of outlet strut fracture for 60 degree Björk-Shiley Convexo-Concave (BSCC) heart valves and new recommendations from an independent expert panel regarding prophylactic valve replacement. The recommendations are described in detail in the enclosed attachments.

Under the Settlement Agreement that was entered into by a worldwide class of BSCC valve patients and Shiley Incorporated and approved by the U.S. District Court in Cincinnati, Ohio in Bowling v. Pfizer, an independent expert medical and scientific panel consisting of cardiothoracic surgeons, cardiologists, and epidemiologists was created called the Supervisory Panel. Under the terms of the Settlement Agreement, the Supervisory Panel is charged with the responsibilities of conducting studies and research, and of making recommendations regarding which BSCC heart valve patients should be considered for prophylactic valve replacement. The Supervisory Panel's recommendations also serve to determine which class members qualify for explantation benefits under the Settlement Agreement. The Panel's work has enabled it to develop these present guidelines for valve replacement surgery. The Panel's guidelines represent a departure from past reliance upon the valve replacement surgery guidelines developed by a Medical Advisory Panel created by Shiley Incorporated. (Letter from the FDA website: http:// www.fda.gov/medwatch/safety/1998/bjork.htm)

In 2004, van Gorp published another study in which he investigated all 2263 Dutch BS convexo-concave patients (van Gorp et al., 2004). For the surviving patients in 1992 ($n = 1330$), they calculated the expected differences in life expectancy with and without valve replacement. Of 1330 patients, 96 (10 percent) had undergone valve replacement. The investigators concluded that 117 patients (9 percent) had an estimated gain in life expectancy after valve replacement.

6.6   Prosthetic Tissue Valves

One disadvantage to a mechanical prosthetic valve is that the patient is required to undergo lifelong anticoagulation therapy. Although pyrolytic carbon has low thrombogenicity, valves made from this material still generate enough blood clots to warrant this type of therapy. Naturally occurring heart valves avoid the blood clotting problem as well as having better hemodynamic properties.

Homografts are valves that come from a different member of the same species. These **allografts** typically come from cadavers and therefore consist of nonliving tissue. The advantage to containing no living cells is that the valves will not be rejected by the immune system of the host. The disadvantage is that they do not have cellular regeneration associated with living tissue and therefore are susceptible to long-term mechanical wear and damage. Human allografts are also available in only very limited supplies. One interesting alternative approach to cadaveric allografts that is sometimes used is to transplant a patient's own pulmonary valve into the aortic position.

One alternative approach to an allograft is a **xenograft**. A xenograft is an implant derived from another species. In 1969, Kaiser and coworkers described a tissue valve substitute that consisted of a gluteraldehyde-treated, explanted porcine valve. This porcine valve became commercially available in 1970 as the Hancock porcine xenograft.

Tissue valves are rarely used in young children and young adults because of valve leaflet calcification and a relatively shorter valve life when compared to mechanical valves.

Review Problems

1. Name four heart valves and describe their location.

2. Describe the composition of a heart valve. Of what materials are they constructed? Do they contain living cells?

3. What driving forces cause the opening and closing of heart valves?

4. In the Özyurda study (2005), the mean pressure gradient across the mitral valve when it was open was 4.0 mmHg for a 31-mm-diameter mitral valve with an effective orifice area of 2.6 cm^2. Estimate the average flow rate through the valve. Is the flow laminar?

5. A patient has a cardiac output of 5.5 L/min. The systolic ejection period is 350 ms with a heart rate of 65 beats per minute. The mean aortic gradient as measured by echocardiography is 75 mmHg. Find the aortic valve area as estimated by the Gorlin equation. What is the average flow rate across the aortic valve during ejection?

6. A particular patient with a stenotic mitral valve has a reduced effective orifice area of the valve: 1.0 cm^2. The systolic ejection period is 330 ms with a heart rate of 70 beats per minute. The mean aortic gradient as measured by echocardiography is 25 mmHg. Using the Gorlin equation, estimate the pressure gradient required to maintain a cardiac output of 4500 mL/min.

7. For an aortic valve with a 0.55 energy loss coefficient, find the effective orifice area based on an aortic cross-sectional area of 5 cm^2.

8. Find a general expression for effective orifice area as a function of loss coefficient.

9. Find the maximum velocity through the vena contracta of the aortic valve for a cardiac output of 6 L/min and an effective orifice area of 0.7 cm^2.

10. The patient from Prob. 5 has a cardiac output of 5.5 L/min. The aortic cross-sectional area is 5 cm^2, blood density is 1060 kg/m^3, and the patient has an effective orifice area of 0.63 cm^2 and a mean aortic gradient of 75 mmHg. Find the energy loss in mmHg and the energy loss coefficient.

11. In general, given the same effective valve area, is a larger aorta associated with a larger or smaller energy loss coefficient? Does that correlate to more or less energy loss?

12. Find the energy loss coefficient for the On-X mitral valve from Prob. 4 described by the Özyurda study (2005). The valve was a 31-mm-diameter mitral valve with an effective orifice area of 2.6 cm^2.

Bibliography

Cannon, SR, Richards, KL, Crawford, M, "Hydraulic Estimation of Stenotic Orifice Area: A Correction of the Gorlin Formula," *Circulation*, 71, 1170–1178, 1985.

Chen, M, Baladi, N, Walji, S, Lam, D, Hanna, E, Lee, G, and Mason, D, Emergency Recognition and Treatment of Strut Fracture and Disc Embolization in Patients with Björk-Shiley Valve Prosthesis. *Am. Heart J.*, December 124(6), 1648–1650, 1992.

de Mol, BA, Kallewaard, M, McLellan, RB, van Herwerden, LA, Defauw, JJ, van der Graaf, Y, "Single-Leg Strut Fractures in Explanted Björk-Shiley Valves," *Lancet*, 343(8888), 9–12, 1994.

Feldman, T, "Percutaneous Valve Repair and Replacement Challenges Encountered, Challenges Met, Challenges Ahead," *Circulation*, 113, 771–773, 2006.

Garcia, D, Pibarot, P, Dumesnil, J. G., Sakr, F., Durand, L-G, "Assessment of Aortic Valve Stenosis Severity: A New Index Based on the Energy Loss Concept," *Circulation*, 101, 765–771, 2000.

Gorlin R, Gorlin SG, "Hydraulic Formula for Calculation of the Area of the Stenotic Mitral Valve, Other Cardiac Valves, and Central Circulatory Shunts," *Am. Heart J.*, 41, 1, 1951.

Hajar, R, "Art and Medicine, Da Vinci's Anatomical Drawings," *Heart*, 3(2), 2002.

Harris, KM, Robiolio, P, "Valvular Heart Disease: Identifying and Managing Mitral and Aortic Lesions," *Post Graduate Medicine On-line*, 106(7), December 1999, http://www.postgradmed.com/issues/1999/12_99/harris.htm.

Ho, S, "Anatomy of the Mitral Valve," *Heart*, 88(Suppl 4), iv5-10, 2002.

Kilner PJ, Yang GZ, Mohiaddin RH, et al., "Helical and Retrograde Secondary Flow Patterns in the Aortic Arch Studied by Three-Directional Magnetic Resonance Velocity Mapping," *Circulation*, 88(I), 2235, 1993.

Kaiser GA, Hancock WD, Lukban SB, Litwak RS. Clinical use of a new design stented xenograft heart valve prosthesis. *Surg. Forum.*, 20, 137–138, 1969.

Lingappa VR and Farey K, *Physiological Medicine*, McGraw-Hill, New York, 2000.

Özyurda, U, Ruchan Akar, A, Uymaz, O, Oguz, M, Ozkan, M, Yildirim, C, Aslan, A, Tasoz, R, "Early Clinical Experience with the On-X Prosthetic Heart Valve," *Interact. Cardiovasc. Thorac. Surg.*, 4, 588–594, 2005; originally published online Sep 16, 2005.

Steyerberg, EW, Kallewaard, M, van der Graaf, Y, van Herwerden, LA, Habbema, JD, "Decision Analyses for Prophylactic Replacement of the Bjork-Shiley Convexo-Concave Heart Valve: An Evaluation of Assumptions and Estimates," *Med. Decis. Making*, 20(1), 20–32, 2000.

Thubrikar, M, Heckman, JL, Nolan, SP, "High Speed Cine-Radiographic Study of Aortic Valve Motion," *J. Heart Valve Dis.*, 2, 653, 1993.

van der Graaf, Y, de Waard, F, van Herwerden, LA, Defauw, J, "Risk of Strut Fracture of Bjork-Shiley Valves," *Lancet*, 339(8788), 257–261, 1992.

van Gorp, MJ, Steyerberg, EW, Van der Graaf, Y, "Decision Guidelines for Prophylactic Replacement of Bjork-Shiley Convexo-Concave Heart Valves: Impact on Clinical Practice," *Circulation*, 109(17), 2092–2096, 2004.

Webb, JG, Chandavimol, M, Thompson, CR, Ricci, DR, Carere, RG, Munt, BI, Buller, CE, Pasupati, S, Lichtenstein, S, "Percutaneous Aortic Valve Implantation Retrograde from the Femoral Artery," *Circulation*, 113, 842–850, 2006; originally published online February 6, 2006.

Westaby, S, Karp, RB, Blackstone, EH, et al., "Adult Human Valve Dimensions and Their Surgical Significance," *Am. J. Cardiol.*, 53, 552, 1984.

Weyman, AE, *Principles and Practices of Echocardiography*, Lea & Febiger, Philadelphia, 1994.

Yoganathan, AP, Hopmeyer, J, Heinrich, RS, in *Biomedical Engineering Handbook*, Bronzino, JD, ed., CRC Press/IEEE Press, Boca Raton, FL, 1995.

Pulsatile Flow in Large Arteries

7.1 Introduction

We now turn our attention to the creation of mathematical models which can be used to simulate the flow of blood in the circulatory system. A good place to start is with the study of flow in large blood vessels. This flow is pulsatile in nature, so our concern is the relationship between the periodically varying pressure gradient driving the flow and the instantaneous blood velocity and flow rate. An understanding of the mathematical description of pulsatile flow will then lead to better experimental procedures as laid out in Chap. 8. This understanding is also the starting point for the more in-depth models of the circulatory system described in Chap. 10.

This chapter is designed as an introduction to two basic modeling techniques. The first of these, due to **J. R. Womersley**, a physiologist at St. Bart's Hospital in London, is based on an analogy with the flow of electricity in a conductor of finite size. You will learn how the spatial distribution of the blood velocity through the artery cross section is described by special mathematical functions called **Bessel functions**. The periodically varying nature of the phenomenon will be captured with the more familiar sines and cosines, which we will refer to as harmonic functions.

The second modeling technique is due to **Joseph C. Greenfield** and **Donald L. Fry**, medical doctors working at the National Heart Institute and Duke University Medical Center in the United States. This approach draws an analog between the blood flow rate and the current flow in a **lumped parameter** electrical circuit having a resistor and an inductor driven with a voltage source. The voltage source is analogous to the pressure gradient driving the pulsatile flow.

In both cases, we will find that the flow rate is harmonic in nature, composed of phase-shifted and scaled sines and cosines. Calculation of the shifting and scaling factors is the key to the solution of the problem. A number of mathematical techniques are required. But, to aid you in getting the big picture, we will try to make this chapter as self-contained as possible, meaning that you should be able to read it without having to use outside reference materials.

So, we ask for your patience and persistence while we take you on a review tour of fluid mechanics notation, the equation of continuity which describes mass conservation in our flow, the use of complex numbers, the practice of representing periodic functions such as pressure gradients with Fourier series of sines and cosines, the basic **Navier–Stokes** equation describing the dynamics of a flowing fluid, and finally concluding with the special solutions by Womersley and Fry and Greenfield.

7.2   Fluid Kinematics

Let's begin with the mathematical description of the motion of the fluid moving in a flow field. It will be convenient to express the velocity in terms of three Cartesian components so that velocity becomes a function of x, y, and z spatial coordinates, as well as a function of time. We also use u, v, and w to stand for the velocity components in the x, y, and z directions, respectively. Written in a concise form:

$$\text{Velocity} = \vec{V}(x,y,z,t)$$

$$\vec{V} = u\hat{i} + v\hat{j} + w\hat{k} = u(x,y,z,t)\hat{i} + v(x,y,z,t)\hat{j} + w(x,y,z,t)\hat{k} \qquad (7.1)$$

In fluid dynamics, a concern is the acceleration of a fluid particle. The corresponding expressions for acceleration are shown in Eqs. (7.2) through (7.5). The chain rule is used in calculating the time derivatives since positions x, y, and z are themselves functions of time. Note that Eq. (7.2) is the vector representation of the acceleration and Eqs. (7.3) through (7.5) show the three scalar Cartesian components of the acceleration.

$$\text{Acceleration} = \vec{a}(x,y,z,t)$$

$$\vec{a} = \frac{\partial \vec{V}}{\partial t} + u\frac{\partial \vec{V}}{\partial x} + v\frac{\partial \vec{V}}{\partial y} + w\frac{\partial \vec{V}}{\partial z} \qquad (7.2)$$

$$a_x = \frac{\partial u}{\partial t} + u\frac{\partial u}{\partial x} + v\frac{\partial u}{\partial y} + w\frac{\partial u}{\partial z} \qquad (7.3)$$

$$a_y = \frac{\partial v}{\partial t} + u\frac{\partial v}{\partial x} + v\frac{\partial v}{\partial y} + w\frac{\partial v}{\partial z} \qquad (7.4)$$

$$a_w = \frac{\partial w}{\partial t} + u\frac{\partial w}{\partial x} + v\frac{\partial w}{\partial y} + w\frac{\partial w}{\partial z} \qquad (7.5)$$

It is also possible to more concisely express the acceleration using the operator $D(\ )/Dt$, which is known as the **material derivative** or **substantial derivative**. The parenthesis in this notation encloses a vector or scalar to be named later, for example, the velocity vector $\vec{V}$ in Eq. (7.6) below.

$$\vec{a} = \frac{D\vec{V}}{Dt} \tag{7.6}$$

The operator $D(\ )/Dt$ is defined in Eq. (7.7).

$$\frac{D(\ )}{Dt} = \frac{\partial(\ )}{\partial t} + u\frac{\partial(\ )}{\partial x} + v\frac{\partial(\ )}{\partial y} + w\frac{\partial(\ )}{\partial z} \tag{7.7}$$

It is valuable here to point out that the total material derivative $\frac{D(\vec{V})}{Dt}$ has two parts. One part is the local time derivative and the other comes from the fact that our fluid particle "convects" or moves into a region of different velocity. Therefore, a particle in steady flow can accelerate. This conception is fundamental to an Eulerian formulation. The derivative is essentially the acceleration of an infinitesimally small chunk of fluid which happens to be located at a point with coordinates (x,y,z) at time denoted by "t."

It is also useful here to define the gradient operator, $\nabla(\ )$, and rewrite the material derivative more concisely.

$$\vec{\nabla}(\ ) = \frac{\partial(\ )}{\partial x}\hat{i} + \frac{\partial(\ )}{\partial y}\hat{j} + \frac{\partial(\ )}{\partial z}\hat{k} \tag{7.8}$$

$$\frac{D(\ )}{Dt} = \frac{\partial(\ )}{\partial t} + \vec{V}\cdot\vec{\nabla}(\ ) \tag{7.9}$$

Equation (7.10) below also introduces the **Laplacian** operator. The Laplacian yields a scalar, by taking the dot product of the vector **gradient operator**, $\vec{\nabla}$, with itself as shown in Eq. (7.11).

Laplacian operator (a scalar)

$$\nabla^2(\ ) = \vec{\nabla}(\ )\cdot\vec{\nabla}(\ ) \tag{7.10}$$

$$\nabla^2(\ ) = \frac{\partial^2(\ )}{\partial x^2} + \frac{\partial^2(\ )}{\partial y^2} + \frac{\partial^2(\ )}{\partial z^2} \tag{7.11}$$

7.3 Continuity

The continuity equation, or conservation of mass, was introduced in Chap. 1 of this book. It was written as $\rho_1 A_1 V_1 = \rho_2 A_2 V_2 = \text{constant}$, where ρ is fluid density, A is the cross-sectional area of the blood vessel, and V is the average blood velocity across the cross section.

Since we are now dealing with the flow at a point the more general version of the conservation of mass equation, which is also commonly referred to as the continuity equation can be written by:

$$\frac{\partial \rho}{\partial t} + \frac{\partial (\rho u)}{\partial x} + \frac{\partial (\rho v)}{\partial y} + \frac{\partial (\rho w)}{\partial z} = 0 \qquad (7.12)$$

The continuity equation is one of the basic governing equations of fluid mechanics and is valid for compressible and incompressible flows in the form written above. It is also valid for steady as well as pulsatile flows in that form.

For the case of incompressible flows, like blood, the continuity equation can also be simplified to:

$$\frac{\partial (u)}{\partial x} + \frac{\partial (v)}{\partial y} + \frac{\partial (w)}{\partial z} = 0 \qquad (7.13)$$

7.4   Complex Numbers

If one uses harmonic functions to model blood flow, then complex numbers are useful. Complex numbers use a number called j (or i in some disciplines), with the definition that $j^2 = -1$. These numbers are written in the form of a vector where $\vec{z} = a + jb$, where a and b are real numbers. The part a is designated the real part of the complex number, $a = \text{Re}(\vec{z})$, and b is the imaginary part, $b = \text{Im}(\vec{z})$. **Leonhard Euler** (1707–1783) made the observation that:

$$e^{jx} = \cos(x) + j\sin(x) \qquad (7.14)$$

Euler's formula became very useful in the solution to differential equations since it allows one to interpret the exponential of a complex number in terms of sines and cosines.

The modern convention is to place the real part of the number on the horizontal x-axis and to display the imaginary part on the y-axis. In this way numbers like $\vec{z} = a + jb$ are displayed in the complex plane. See Fig. 7.1, which shows a vector $\vec{z}$ represented in the complex plane.

In this presentation, we will use Euler's identity in the following form:

$$e^{j\omega t} = \cos(\omega t) + j\sin(\omega t) \qquad (7.15)$$

where $j^2 = -1, j^3 = -j$, t = time, ω = fundamental frequency of the signal, $\cos(\omega t) = \text{Re}[e^{j\omega t}]$ or the real part, and $\sin(\omega t) = \text{Im}[e^{j\omega t}]$ or the imaginary part

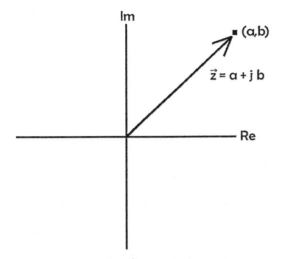

Figure 7.1 The vector $\vec{z} = a + jb$ in the complex plane.

In the type of fluid mechanics problem with which we are dealing in this book, t represents the independent variable, time. There is also a fundamental frequency, ω, associated with the problem, which is typically the heart rate in radians per second. It is also possible to represent (ωt) as θ, now with units of radians. If the heart beat takes place over a period of T_0 seconds, the fundamental frequency in radians per second is 2π radians divided by T_0 seconds. The fundamental frequency in hertz is $1/T_0$, which is not to be used with our harmonic functions.

Now that we have this useful concept of complex numbers, it is possible to add, subtract, multiply, and divide the numbers. For example, to add complex numbers it is necessary to add the real parts and the imaginary parts separately.

$$(A + Bj) + (C + Dj) = (A + C) + (B + D)j$$

For multiplication it is necessary to convert to the exponential form with a magnitude and an angle and to multiply the magnitudes and add the angles.

$$\vec{A}\,\vec{B} = (Ae^{j\theta})(Be^{j\phi}) = ABe^{j(\theta+\phi)}$$

The complex conjugate of $\vec{z} = a + jb$ is defined as $\vec{z}^{\,*} = a - jb$. The square root of a complex number $\vec{z}$ times its complex conjugate results in the magnitude of that number. The magnitude of $\vec{z}$ is $\sqrt{a^2 + b^2}$ which could also be written as $\sqrt{(\vec{z}) \cdot (\vec{z}^{\,*})}$. The argument of $\vec{z}$ is $\tan^{-1}(b/a)$.

Example Problem Using Complex Numbers   Here is a typical complex number, expressed in "rectangular" form using real and imaginary parts.

$$\vec{A} = (3 + 4j)$$

The magnitude of $\vec{A}$ is written $|\vec{A}| = \sqrt{3^2 + 4^2} = 5$

The angle theta is written $\theta = \tan^{-1}\left(\dfrac{4}{3}\right) = 53.1°$

So an alternate form of writing the complex number in phasor notation is $\vec{A} = 5\angle 53.1°$.

A second example of a complex number is $\vec{B} = 5 + 12j$

now the magnitude of $\vec{B}$ is $|\vec{B}| = \sqrt{5^2 + 12^2} = 13$

and the angle phi is $\phi = \tan^{-1}\left(\dfrac{12}{5}\right) = 67.4°$

Finally the phasor notation is $\vec{B} = 13 \angle 67.4°$

Now from our two complex number examples, it is possible to show a multiplication example:

$$\vec{A}\,\vec{B} = (5e^{j53.1})\,(13e^{j67.4}) = 65e^{j(120.5)}$$

$$65e^{j(120.5)} = 65[\cos(120.5) + j\sin(120.5)] = -33 + 56j$$

From the complex number $\vec{A}$ and $\vec{B}$ the division example is written:

$$\vec{A}/\vec{B} = (5e^{j53.1})/(13e^{j67.4}) = \frac{5}{13}e^{j(53.1-67.4)} = \frac{5}{13}e^{j(-14.3)}$$

$$= \frac{5}{13}[\cos(-14.3) + j\sin(-14.3)]$$

$$= 0.373 + -0.0947j$$

7.5 Fourier Series Representation

The **Fourier series** theorem which was published in 1822 states *"A periodic function f(t) with a period T can be represented by the sum of a constant term, a fundamental of period T, and its harmonics."* In general, a Fourier series can be used to expand any periodic function into an infinite sum of sines and cosines. This is an extremely useful way to break up an arbitrary periodic function into a set of simple terms.

For the case of pulsatile flow, where u is velocity and P is pressure, the partial derivative of velocity u with respect to time is not zero. In addition the partial derivative of pressure P with respect to distance along the tube x is also nonzero. These were important conditions for Poiseuille's law. Therefore, Poiseuille flow is no longer a good estimate for the case of pulsatile flow.

Written more concisely:

$$\frac{\partial u}{\partial t} \neq 0 \quad \text{and} \quad \frac{\partial P}{\partial x} \neq 0$$

Since the pressure waveform is periodic, it is convenient to write the partial derivative of pressure $\partial P/\partial x$ using a Fourier series. Figure 7.2 shows a plot of a pulsatile pressure waveform that shows pressure drop versus time for a pulsatile flow condition.

This periodic function depends on the fundamental frequency of the signal ω (heart rate in rad/s) and the time t. We can write any such function as a sum of sine and cosine terms, with appropriate coefficients, known as Fourier coefficients.

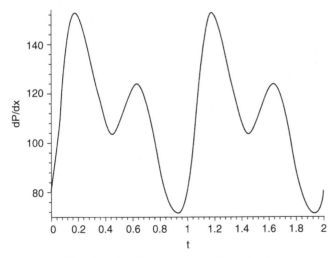

Figure 7.2   Plot of a pulsatile pressure waveform showing pressure drop versus time for a pulsatile flow condition.

Since $\frac{\partial P}{\partial x}$ is periodic, it may be written:

$$\frac{\partial P}{\partial x} \approx A_o + A_1 \cos(\omega t) + A_2 \cos(2\omega t) + A_3 \cos(3\omega t) + \cdots$$

$$+ B_1 \sin(\omega t) + B_2 \sin(2\omega t) + B_3 \sin(3\omega t) + \cdots$$

This is a Fourier series representation of the pressure gradient. If we have a periodic pressure waveform, it is possible to obtain the Fourier coefficients, $A_1, A_2, A_3, \ldots B_1, B_2, B_3, \ldots$ by a number of common methods. The formal approach taught in most courses in differential equations involves the evaluation of a series of integrals. Let f be a periodic function with period T_0. Then

$$A_0 = \frac{1}{T_0} \int_0^{T_0} f(t)\, dt \tag{7.17}$$

$$A_n = \frac{2}{T_0} \int_0^{T_0} f(t) \cos(n\omega t)\, dt \tag{7.18}$$

$$B_n = \frac{2}{T_0} \int_0^{T_0} f(t) \sin(n\omega t)\, dt \tag{7.19}$$

Doing the integrals formally using the principles of integral calculus can be very laborious if f is other than a fairly simple function. It may be convenient to evaluate the integrals numerically. Or, if a set of digitized data exists, for example, a fast Fourier transform implemented in many types of software may be used to approximate the needed Fourier coefficients.

It is also possible to write the Fourier series using complex numbers by using Euler's identity to convert the sines and cosines to exponentials. The equivalent expression would be:

$$\frac{\partial P}{\partial x} = \text{Re}\left[\sum_{n=0}^{\infty} a_n e^{j\omega n t} \right] \tag{7.20}$$

where $a_n = A_n - B_n j$, n is the number of the harmonic, ω is the fundamental frequency (rad/s), and A_o is the mean pressure gradient (nonpulsatile).

Note:
$$a_n = (An - B_n j)$$

or

$$a_n = \text{conjugate}\,(A_n + B_n j)$$

Example Problem with Fourier Series Laboratory measurements on the aorta of a pig determine the pressure gradient shown in Fig. 7.3. What we see is a periodic function with a period of 1.0 s.

Calculate the Fourier coefficients of the waveform. That means that we want to find numbers $A_0, A_1, \ldots, A_n$ and $B_1, B_2, \ldots, B_n$ so that we can describe the waveform with the following equation.

$$\frac{dP}{dx}(t) = A_0 + \sum_{n=1}^{N} A_n \cos(n\omega t) + \sum_{n=1}^{N} B_n \sin(n\omega t)$$

The fundamental frequency ω is calculated using the period of the waveform T_0.

$$\omega = \frac{2\pi}{T_0}$$

Solution: The fundamental frequency is 1.0 Hz, which we convert to radians per second. The period T_0 is equal to 1.0 s. Therefore,

$$\omega = 2\pi \text{ rad/s} = 6.283 \text{ rad/s}$$

Note that the function in the first 0.4 s can be described with

$$\frac{dP}{dx}(t) = -3000 + \frac{3000}{0.4}t = -3000 + 7500t$$

and the function is zero in the final 0.6 s. We start with the first term in the Fourier series, often referred to as the DC term,

$$a_0 = \frac{1}{T_0} \int_0^{T_0} \frac{dP}{dx}(t)\, dt = \int_0^{0.4} (-3000 + 7500t)\, dt = 600$$

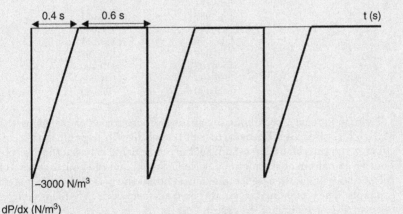

Figure 7.3 Schematic of pressure gradient in an artery.

Further cosine terms are calculated using the following formulas.

$$A_n = \frac{2}{T_0}\int_0^{T_0}\frac{dP}{dx}(t)\cos(n\omega t)\,dt = \int_0^{0.4}(-3000 + 7500t)\cos(n\omega t)\,dt$$

$$A_n = \frac{379.95\,(\cos(2.5133n) - 1)}{n^2}$$

These values can be calculated by evaluating the integral using calculus as shown or else by numerical integration in a program such as Matlab. The table contains the first 10 values of this coefficient. We also need coefficients for the sine terms.

$$B_n = \frac{2}{T_0}\int_0^{T_0}\frac{dP}{dx}(t)\sin(n\omega t)\,dt = \int_0^{0.4}(-3000 + 7500t)\sin(n\omega t)\,dt$$

$$B_n = \frac{75.991\,(5\sin(2.5133n) - 12.566)}{n^2}$$

These values are also calculated and presented in the table. We suggest that you just try to check one or two of these numbers using the formulas above or with numerical integration. The Fourier coefficients in the table were calculated with the Matlab program. Checking against the calculus formulas reveals no differences in the digits presented.

Table of Fourier Coefficients for Example Problem

n	A_n	B_n
0	−600	—
1	−687.344	−731.598
2	−65.636	−567.804
3	−29.171	−278.159
4	−42.960	−252.691
5	0.000	−190.986
6	−19.093	−152.951
7	−5.358	−143.793
8	−4.102	−113.720
9	−8.486	−108.860
10	0.000	−95.493

There is an easier technique to calculate Fourier coefficients using the Matlab **Fast Fourier Transform** (fft) capability. To accomplish this calculation requires an understanding of the relationship between the discrete Fourier transform of samples of a periodic signal and the Fourier series of the periodic signal. In essence, one must choose one period of the signal as a sample. Then, one must calculate the transform with fft, followed by a rearranging or scaling of the result.

Suppose that n is the number of points sampled. The fft works optimally if n is a power of 2 such as 512 or 1024. Let $N = n/2$. Next, assume that the actual sample is contained in the n-vector, y. The following Matlab commands are needed.

```
Compute the transform
Y = fft(y);
% Rearrange values.(Negative harmonics considered here!)
Y = [ conj(Y(N+1)) Y(N+2:end) Y(1:N+1)];
% Scale values by n
Y = Y / n;
% Get DC term
a0 = Y(N+1);
% Calculate cosine terms
an = 2*real(Y(N+2:end));
% Calculate sine terms
bn = -2*imag(Y(N+2:end));
```

At this point the variable a0, and the arrays an and bn, will contain the coefficients. Unfortunately, due to the finite sampling, the higher-order terms are increasingly less accurate. So we would want to have n to be much greater than the number of Fourier terms that we actually propose to use in order to take this approach. We will now repeat the table above, including the terms calculated with the fft method in order that they may be compared. In this case $n = 1024$ and we are proposing to work with the first 10 harmonics. The results may be seen to be close.

n	A_n	An - fft	B_n	Bn - fft
0	−600	−600.879	—	—
1	−687.344	−690.397	−731.598	−730.622
2	−65.636	−69.0748	−567.804	−568.03
3	−29.171	−31.8351	−278.159	−278.208
4	−42.960	−46.0682	−252.691	−252.503
5	0.000	−2.93198	−190.986	−191.158
6	−19.093	−21.9501	−152.951	−152.802
7	−5.358	−8.42129	−143.793	−143.82
8	−4.102	−6.92647	−113.720	−113.728
9	−8.486	−11.485	−108.860	−108.752
10	0.000	−2.93198	−95.493	−95.5564

Finally, we plot the 10-term Fourier series result, comparing it with the function that it represents. See Fig. 7.4.

Notice how close the Fourier series is to the function during most of the period. Only near the discontinuity which occurs at the beginning and end of the period there is noticeable waviness by the Fourier series as it attempts to converge on a point midway between 0 and −3000. As more terms are added, this waviness shifts closer and closer to the discontinuity, finally being observed as a kind of spike. This type of behavior is known as Gibbs' phenomenon.

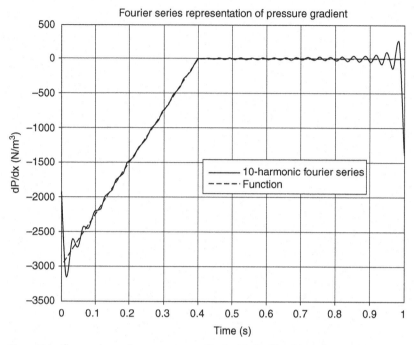

Figure 7.4 Comparison of a pressure gradient with its Fourier series.

7.6   Navier–Stokes Equations

The Navier–Stokes equations are named in honor of the French mathematician L. M. H. Navier (1758–1836) and the English mechanician Sir G. G. Stokes (1819–1903), who were responsible for their formulation. These equations are nonlinear, second-order, partial differential equations and they are considered to be the governing differential equations of motion for incompressible, Newtonian fluids.

The Navier–Stokes equations can now be written rather efficiently in the following form:

$$\rho \frac{D \vec{V}}{Dt} = \rho \vec{g} - \vec{\nabla} P + \mu \nabla^2 \vec{V} \qquad (7.21)$$

In Eq. (7.21), ρ is the fluid density, DV/Dt is the material derivative of the velocity vector, $\vec{g}$ is gravitational acceleration, $\vec{\nabla} P$ represents the pressure gradient, and μ represents viscosity.

The left-hand side of Eq. (7.21), $\rho \frac{D \vec{V}}{Dt}$, represents something analogous to a mass (density, ρ) times an acceleration (substantial derivative of the velocity vector) and can be expanded to the following three scalar equations:

$$\rho\,a_x = \underbrace{\rho\,\frac{\partial(u)}{\partial t}}_{\substack{\text{transient}\\\text{inertia term}}} + \rho\left(\overbrace{u\frac{\partial(u)}{\partial x}}^{\substack{\text{nonlinear in depend-}\\\text{ent variable}}} + v\frac{\partial(u)}{\partial y} + w\frac{\partial(u)}{\partial z} \right)$$

$$\underbrace{\qquad\qquad\qquad}_{\text{convective inertia term}}$$

(7.22)

$$\rho\,a_y = \underbrace{\rho\,\frac{\partial(v)}{\partial t}}_{\substack{\text{transient}\\\text{inertia term}}} + \rho\left(u\frac{\partial(v)}{\partial x} + \overbrace{v\frac{\partial(v)}{\partial y}}^{\substack{\text{nonlinear in depend-}\\\text{ent variable}}} + w\frac{\partial(v)}{\partial z} \right)$$

$$\underbrace{\qquad\qquad\qquad}_{\text{convective inertia term}}$$

(7.23)

$$\rho\,a_z = \underbrace{\rho\,\frac{\partial(w)}{\partial t}}_{\substack{\text{transient}\\\text{inertia term}}} + \rho\left(u\frac{\partial(w)}{\partial x} + v\frac{\partial(w)}{\partial y} + \overbrace{w\frac{\partial(w)}{\partial z}}^{\substack{\text{nonlinear in}\\\text{dependent variable}}} \right)$$

$$\underbrace{\qquad\qquad\qquad}_{\text{convective inertia term}}$$

(7.24)

Note that Eqs. (7.22) through (7.24) are formulated on a per volume basis. The reader can observe, for example, that $\frac{\partial P}{\partial x}$ has units of $\frac{F/L^2}{L} = \frac{F}{L^3} = \frac{Force}{Volume}$. One might wonder what kinds of forces are causing all of that mass to accelerate. The right-hand side of the Navier–Stokes equations provides insight into that question. The right-hand side can be expanded into the three scalar expressions shown below.

$$\underbrace{\rho g_x}_{\text{weight}} - \underbrace{\frac{\partial P}{\partial x}}_{\text{pressure force}} + \mu\left(\frac{\partial^2(u)}{\partial x^2} + \frac{\partial^2(u)}{\partial y^2} + \overbrace{\frac{\partial^2(u)}{\partial z^2}}^{\substack{\text{higher ordered deriv-}\\\text{ative, 2nd-order PDE}}} \right)$$

$$\underbrace{\qquad\qquad\qquad}_{\text{viscous forces}}$$

(7.25)

$$\rho g_y \underbrace{}_{\text{weight}} - \underbrace{\frac{\partial P}{\partial y}}_{\text{pressure force}} + \mu \left(\underbrace{\frac{\partial^2(v)}{\partial x^2} + \frac{\partial^2(v)}{\partial y^2} + \overbrace{\frac{\partial^2(v)}{\partial z^2}}^{\substack{\text{higher ordered deriv-} \\ \text{ative, 2nd-order PDE}}}}_{\text{viscous forces}} \right) \tag{7.26}$$

$$\rho g_z \underbrace{}_{\text{weight}} - \underbrace{\frac{\partial P}{\partial z}}_{\text{pressure force}} + \mu \left(\underbrace{\frac{\partial^2(w)}{\partial x^2} + \frac{\partial^2(w)}{\partial y^2} + \overbrace{\frac{\partial^2(w)}{\partial z^2}}^{\substack{\text{higher ordered deriv-} \\ \text{ative, 2nd-order PDE}}}}_{\text{viscous forces}} \right) \tag{7.27}$$

The left-hand side of the equation could also be written in terms of cylindrical components. For blood flow problems in which the blood is flowing in cylindrical arteries, this type of coordinate system is convenient. The expression for the x-components of inertia (axial direction) using cylindrical components is shown below.

$$\rho \underbrace{\frac{\partial(u)}{\partial t}}_{\substack{\text{transient} \\ \text{inertia term}}} + \rho \left(\underbrace{\overbrace{u\frac{\partial(u)}{\partial x}}^{\substack{\text{nonlinear in depe-} \\ \text{ndent variable}}} + v_r \frac{\partial(u)}{\partial r} + \frac{v_\theta}{r} \frac{\partial(u)}{\partial \theta}}_{\text{convective inertia term}} \right) \tag{7.28}$$

The right-hand side of the Navier–Stokes equation written in cylindrical components for the axial component then completes the equation:

$$\rho \underbrace{\frac{\partial(u)}{\partial t}}_{\substack{\text{transient} \\ \text{inertia term}}} + \rho \left(\underbrace{\overbrace{u\frac{\partial(u)}{\partial x}}^{\substack{\text{nonlinear in depe-} \\ \text{ndent variable}}} + v_r \frac{\partial(u)}{\partial r} + \frac{v_\theta}{r} \frac{\partial(u)}{\partial \theta}}_{\text{convective inertia term}} \right)$$

$$= \rho g_x - \frac{\partial P}{\partial x} + \mu \left(\frac{\partial^2(u)}{\partial x^2} + \frac{\partial^2(u)}{\partial r^2} + \frac{1}{r} \frac{\partial(u)}{\partial r} + \frac{1}{r^2} \frac{\partial^2(u)}{\partial \theta^2} \right) \tag{7.29}$$

Example Problem Using Navier–Stokes Show that the parabolic velocity profile from horizontal Poiseuille flow is a solution of the Navier–Stokes equations using the cylindrical form.

Solution: For Poiseuille flow we have already assumed:

1. Steady flow
2. Uniform flow (constant cross section)
3. Laminar flow
4. Axially symmetric pipe flow

Therefore, there is no change in velocity with respect to time, so $\frac{\partial u}{\partial t} = 0$. Since the flow is axially symmetric, there is no swirling flow and no velocity in the radial or transverse direction and therefore no change in velocity in either the radial or transverse direction, so $\frac{\partial u}{\partial \theta} = 0$ and $v_r = 0$ and $v_\theta = 0$. Since Poiseuille flow is uniform, there is no change in velocity u in the x (axial) direction, so $\frac{\partial u}{\partial x} = 0$. From Eq. (1.11) in Chap. 1, we saw that the parabolic velocity profile from Poiseuille flow has the form

$$u = \frac{1}{4\mu} \frac{dP}{dx} [r^2 - R^2]$$

where u is the velocity component along the tube as a function of r, μ is fluid viscosity (a constant for Newtonian fluids), $\frac{\partial P}{\partial x}$ is the pressure gradient along the tube (a constant for Poiseuille flow), r is the radius variable, and R is the radius of the tube (a constant).

The Navier–Stokes equation for the component along the axial direction can be written from Eqs. (7.28) and (7.29).

$$\rho \frac{\partial(u)}{\partial t} + \rho\left(u \frac{\partial(u)}{\partial x} + v_r \frac{\partial(u)}{\partial r} + \frac{v_\theta}{r} \frac{\partial(u)}{\partial \theta}\right)$$

$$= \rho g_x - \frac{\partial P}{\partial x} + \mu \left(\frac{\partial^2(u)}{\partial x^2} + \frac{\partial^2(u)}{\partial r^2} + \frac{1}{r} \frac{\partial(u)}{\partial r} + \frac{1}{r^2} \frac{\partial^2(u)}{\partial \theta^2}\right)$$

(7.30)

The equation has now become somewhat more simplified. Equation (7.31) tells us that for Poiseuille flow, there are no accelerations and that viscous forces must balance the pressure forces and gravitational forces.

$$0 = \rho g_x - \frac{\partial P}{\partial x} + \mu\left(\frac{\partial^2(u)}{\partial x^2} + \frac{\partial^2(u)}{\partial r^2} + \frac{1}{r} \frac{\partial(u)}{\partial r} + \frac{1}{r^2} \frac{\partial^2(u)}{\partial \theta^2}\right) \qquad (7.31)$$

Since this flow is also horizontal and as already mentioned u does not vary in the x direction or the θ direction, Eq. (7.31) further simplifies to:

$$\frac{\partial P}{\partial x} = \mu\left(\frac{\partial^2(u)}{\partial r^2} + \frac{1}{r} \frac{\partial(u)}{\partial r}\right) \qquad (7.32)$$

Since the proposed solution to the differential equation in Eq. (7.32) is

$$u = \frac{1}{4\mu} \frac{dP}{dx} [r^2 - R^2]$$

then we need to find the first and second derivatives of u with respect to r and plug it into the equation to check for equality. We should also recognize explicitly that $\frac{dP}{dx} = \frac{\partial P}{\partial x}$ since the pressure under Poiseuille flow conditions is only a function of x and does not vary with time.

$$\frac{\partial u}{\partial r} = \frac{2}{4\mu}\frac{dP}{dx}r \tag{7.33}$$

$$\frac{\partial^2 u}{\partial r^2} = \frac{2}{4\mu}\frac{dP}{dx} \tag{7.34}$$

$$\frac{\partial P}{\partial x} = \mu\left(\left(\frac{2}{4\mu}\frac{dP}{dx}\right) + \frac{1}{r}\left(\frac{2}{4\mu}\frac{dP}{dx}r\right)\right) \tag{7.35}$$

The equality holds true showing that the Poiseuille flow parabolic velocity profile is one simple solution of the Navier–Stokes equations.

7.7 Pulsatile Flow in Rigid Tubes—Womersley Solution

Consider the following problem. Given a pressure waveform and the geometry of the artery and some fluid characteristics, estimate the flow rate in terms of volume per time. This may be a pulsatile pressure waveform as is seen in the relatively larger arteries in humans. The blood flow rate is a measure of perfusion, or simply said, how much oxygen may be provided to the downstream, or distal tissues. Therefore, a mathematical model of flow rate under pulsatile conditions can be a very useful device.

To estimate flow from a pulsatile driving pressure in rigid tubes, we will begin by assuming a Newtonian fluid, uniform, laminar, axially symmetric, pipe flow. This is similar to the Poiseuille flow problem, but now we are considering pulsatile flow rather than steady.

We will need to have a driving function for the pressure. Recall from Eq. (7.20) that

$$\frac{\partial P}{\partial x} = \text{Re}\left[\sum_{n=0}^{\infty} a_n e^{j\omega nt}\right] \tag{7.20}$$

Therefore, for each harmonic n, we can write each component of the driving pressure as a complex exponential by using Eq. (7.36).

$$\frac{\partial P}{\partial x}\bigg|_n = a_n e^{j\omega nt} \tag{7.36}$$

For each component of the driving pressure from 0 to n, it is also possible to write the Navier–Stokes equation given in Eq. (7.30). The use of the subscript "n" emphasizes that we are solving for one generic "nth" component and then, at the end, assembling the total solution as a sum of components, and, finally, taking the real part.

$$
\rho \frac{\partial(u)}{\partial t} + \rho \left(u\frac{\partial(u)}{\partial x} + v_r\frac{\partial(u)}{\partial r} + \frac{v_\theta}{r}\frac{\partial(u)}{\partial \theta} \right)
$$

$$
= \rho g_x - \frac{\partial P}{\partial x} + \mu \left(\frac{\partial^2(u)}{\partial x^2} + \frac{\partial^2(u)}{\partial r^2} + \frac{1}{r}\frac{\partial(u)}{\partial r} + \frac{1}{r^2}\frac{\partial^2(u)}{\partial \theta^2} \right)
$$

(7.30)

For this pulsatile flow case we have assumed steady, uniform, laminar, axially symmetric pipe flow. Since the flow is axially symmetric, there is no swirling flow and no velocity in the radial or transverse directions and therefore no change in velocity in either the radial or transverse direction, so $\frac{\partial u}{\partial \theta} = 0$ and $v_r = 0$ and $v_\theta = 0$. Since the flow is uniform, there is no change in velocity u in the x (axial) direction, so $\frac{\partial u}{\partial x} = 0$. The flow is also horizontal; therefore, Eq. (7.30) simplifies to the complex equation:

$$
\frac{a_n e^{j\omega n t}}{\rho} = \nu \frac{\partial^2(u_n)}{\partial r^2} + \frac{\nu}{r}\frac{\partial(u_n)}{\partial r} - \frac{\partial(u_n)}{\partial t}
$$

(7.37)

This is a linear, second-order, partial differential equation (PDE) with a driving function. One possible solution to the PDE is:

$$
u_n = \underbrace{f_n(r)}_{\substack{f \text{ is not time} \\ \text{dependent}}} e^{j\omega n t}
$$

(7.38)

In order to try this solution in Eq. (7.37), we will need to have the derivative of u with respect to time t, the derivative of u with respect to r, and also the second derivative of u with respect to r.

$$
\frac{du_n}{dr} = \frac{df_n(r)}{dr} e^{j\omega n t}
$$

(7.39)

$$
\frac{d^2 u_n}{dr^2} = \frac{d^2 f_n(r)}{dr^2} e^{j\omega n t}
$$

(7.40)

$$
\frac{du_n}{dt} = j\omega n f_n(r) e^{j\omega n t}
$$

(7.41)

Now by using our proposed solution to the differential equation, it is possible to rewrite Eq. (7.37) in the following manner so that an exponential appears in each term.

$$\frac{a_n e^{j\omega nt}}{\rho} = v \frac{d^2 f_n(r)}{dr^2} e^{j\omega nt} + \frac{v}{r} \frac{df_n(r)}{dr} e^{j\omega nt} - j\omega n f_n(r) e^{j\omega nt} \qquad (7.42)$$

Now it is possible to divide Eq. (7.42) by $e^{j\omega nt}$ to remove the time dependence. The result would be:

$$\frac{a_n}{\rho} = v \frac{d^2 f_n(r)}{dr^2} + \frac{v}{r} \frac{df_n(r)}{dr} - j\omega n f_n(r) \qquad (7.43)$$

This allows us to treat time dependence and spatial dependence separately. Also we can replace v with μ/ρ and $-j$ with j^3 to obtain Eq. (7.44).

$$\frac{a_n}{\mu} = \frac{d^2 f_n(r)}{dr^2} + \frac{1}{r} \frac{df_n(r)}{dr} + \frac{j^3 \omega n f_n(r)}{v} \qquad (7.44)$$

Now we have an ordinary differential equation instead of a partial differential equation and one that is not time dependent! Equation (7.44) looks a little bit like a zero-order Bessel differential equation. The homogeneous differential equation for the case without a driving function would be:

$$\frac{d^2 f_n(r)}{dr^2} + \frac{1}{r} \frac{df_n(r)}{dr} + \frac{j^3 \omega n f_n(r)}{v} = 0 \qquad (7.45)$$

This equation matches a template. A zero-order Bessel differential equation has the form:

$$\frac{d^2 g}{ds^2} + \frac{1}{s} \frac{dg}{ds} + \lambda^2 g = 0 \qquad (7.46)$$

The solution to Eq. (7.46) is a Bessel function of order zero and complex arguments, which is well known and arises in problems connected with the distribution of current in conductors of finite size. One homogenous solution is:

$$f_n(r) = C_1 J_0(\lambda r) + C_2 Y_0(\lambda r) \qquad (7.47)$$

where C_1 and C_2 are constants, J_0 is a zero-order Bessel function of the first kind, Y_0 is a zero-order Bessel function of the second kind, λ is defined in Eq. (7.48), and J_0 is given by Eq. (7.49). It can easily be shown, if you know Bessel functions, that in this case $C_2 = 0$, and we need only to be concerned with the first term, the one having J_0. By comparing our equation with its Bessel equation template, we can write

$$\lambda^2 = \frac{j^3 \omega n}{\nu} \tag{7.48}$$

$$J_o(x) = 1 - \frac{x^2}{2^2(1!)^2} + \frac{x^4}{2^4(2!)^2} - \frac{x^6}{2^6(3!)^2} + \frac{x^8}{2^8(4!)^2} - \cdots \tag{7.49}$$

Note that λ^2 is a constant for a given kinematic viscosity, fundamental frequency, and harmonic n.

The Bessel equation in the Womersley solution is inhomogeneous because of the pressure gradient term. We need to find a find a particular solution to the differential equation, so we propose an easy one.

Let
$$f_n(r)_{\text{particular}} = C_3$$

We can then set the derivatives of $f_n(r)$ with respect to r to zero and solve for the constant C_3.

$$\frac{j^3 \omega n}{\nu} C_3 = \frac{a_n}{\mu} \tag{7.50}$$

$$C_3 = \frac{a_n}{\mu} \frac{\mu}{\rho} \frac{1}{j^3 \omega n} = \frac{-a_n}{j\rho\omega n} \tag{7.51}$$

The total solution becomes:

$$f_n(r) = C_1 J_o(\lambda r) - \frac{a_n}{j\rho\omega n} \tag{7.52}$$

According to the no-slip condition, velocity at the wall should be zero. That location is also defined by $r = R$, since $r = 0$ corresponds to the centerline of the vessel. In order to solve for the constant C_1, use the boundary condition of $u = 0$ at $r = R$.

$$0 = C_1 J_o(\lambda R) - \frac{a_n}{j\rho\omega n} \tag{7.53}$$

$$C_1 = \frac{a_n}{j\rho\omega n J_o(\lambda R)} \tag{7.54}$$

The total solution to the differential equation, for harmonic n, now becomes:

$$f_n(r) = \frac{a_n}{j\rho\omega n} \left[\frac{J_o(\lambda r)}{J_o(\lambda R)} - 1 \right] \tag{7.55}$$

This is still the complex form that is time independent. If we take the pressure gradient as the real part of $a_n e^{j\omega nt}$ and substitute u from Eq. (7.38), the corresponding velocity as a function of r and t becomes:

$$u_n = \text{Re}\left[\frac{a_n}{j\rho\omega n}\left\{\frac{J_0(\lambda r)}{J_0(\lambda R)} - 1\right\}e^{j\omega nt}\right] \tag{7.56}$$

Recall that this solution applies to the results of each harmonic. Now to find the velocity as a function of radius r and time t for the entire driving pressure, add together steady flow result u_0 to the results from all harmonics.

$$u(r,t) = u_0(r) + \sum_{n=1}^{\infty} u_n(r,t) \tag{7.57}$$

Figure 7.5 shows an example plot of the total velocity $u(r)$ for two different r values in a mathematical model of an artery under pulsatile driving pressure.

At this point we turn to a quantity that is a bit more important than the velocity. It is the flow rate passing through a given cross section of the blood vessel. To find this flow rate one needs only to integrate the velocity function just found multiplied by the differential area, over the entire cross section. The differential area, a simple annulus, may be written as a function of r, that is, $2\pi r dr$, so the flow term becomes

$$Q(t) = \int_0^R u(r,t) \cdot 2\pi r\, dr \tag{7.58}$$

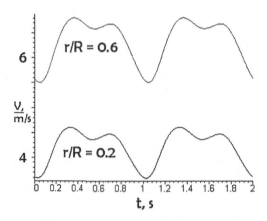

Figure 7.5 Graph of axial velocity as a function of radius and time, where r is the radius variable and R is the radius of the artery.

We can perform this integral after looking up integral identities for Bessel functions in a mathematical reference. The one we need to use is

$$\int x J_0(x)\,dx = x J_1(x)$$

The symbol J_1 denotes a Bessel function of the first kind and of the first order. After a number of steps of calculus style derivations which need not be explained here, we reach a solution for the flow rate produced by harmonic n of the pressure gradient.

$$Q_n = ((\pi R^2))\text{Re}\left[\frac{a_n}{j\omega n\rho}\left\{\frac{2 J_1(\lambda R)}{\lambda R J_0(\lambda R)} - 1\right\}e^{j\omega nt}\right] \qquad (7.59)$$

These must be summed and added to

$$Q_0 = a_0\frac{8\mu}{\pi R^4} \qquad (7.60)$$

which is the average flow rate produced by the constant term in the pressure gradient's Fourier series. Finally, we have

$$Q(t) = Q_0 + \sum_{n=1}^{N} Q_n(t) \qquad (7.61)$$

Womersley published his solution in 1955 when it wasn't so easy to make up a computer model of the flow. Bessel functions were not easily accessible then as they are now in all scientific software. So he integrated the velocity terms, solved for flow, and published the following equation for the flow component resulting from each of the driving pressure gradient harmonics:

$$Q_n = \frac{\pi R^4}{\mu} M_n \left(\frac{M_{10}}{\alpha^2}\right)_n \sin\left(\omega nt + \phi_n + \varepsilon_{10_n}\right) \qquad (7.62)$$

where the pressure gradient associated with each harmonic is:

$$\frac{\partial P}{\partial x}\bigg|_n = M_n \cos\left(\omega nt + \phi_n\right) \qquad (7.63)$$

The magnitude of the driving pressure is given by:

$$M_n = \sqrt{A_n^2 + B_n^2} \qquad (7.64)$$

or alternatively, recall that $a_n = A_n - B_n j$

$$M_n = |a_n| \qquad (7.65)$$

The angle of the phase shift is given by:

$$\phi_n = \text{argument}(-a_n) \qquad (7.66)$$

The total flow Q is computed with exactly the same kind of sum just described.

Womersley compiled the values of the constants for $\frac{M_{10}}{\alpha^2}$ and for ε_{10} as a function of the α parameter. As α approaches zero, $\frac{M_{10}}{\alpha^2}$ approaches 1/8 and ε_{10} approaches 90° and the solution becomes Poiseuille's law. The reader may wish to use this alternative to the formula for flow rate involving Bessel functions since some spreadsheets do not provide Bessel functions with complex arguments.

For the sake of completeness, the calculation of M_{10} and ε_{10} will now be briefly described. For each harmonic, start by finding

$$M_0 = |J_0(\alpha j^{3/2})| \quad \text{and} \quad M_1 = |J_1(\alpha j^{3/2})|$$

$$\theta_0 = \text{argument}(J_0(\alpha j^{3/2})) \quad \text{and} \quad \theta_1 = \text{argument}(J_1(\alpha j^{3/2}))$$

(The number of the harmonic enters through α.) Next let

$$\delta_{10} = \frac{3}{4}\pi - \theta_1 + \theta_0$$

and

$$k = \frac{\alpha M_0}{2 M_1}$$

The quantities of interest may now be found.

$$M_{10} = \frac{1}{k}\sqrt{(\sin(\delta_{10}))^2 + (k - \cos(\delta_{10}))^2} \qquad (7.67)$$

$$\varepsilon_{10} = \tan^{-1}\left(\frac{\sin(\delta_{10})}{k - \cos(\delta_{10})}\right) \qquad (7.68)$$

Recall from Chap. 1 that α is the Womersley number, or alpha parameter, which is a ratio of transient to viscous forces, and is defined by:

$$\alpha = r\sqrt{\frac{\omega\rho}{\mu}} \qquad (7.69)$$

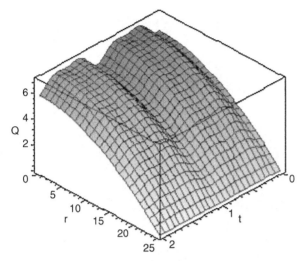

Figure 7.6 Three-dimensional plot that shows a flow wave-form plotted in terms of time and vessel radius for a pulsatile flow condition. The curve was generated using Womersley's version of the Navier–Stokes equation solution. The flow Q is shown in cubic millimeters per second, the time t is shown in seconds, the radius r in millimeters.

Figure 7.6 shows a velocity waveform plotted in terms of time and vessel radius for a pulsatile flow condition in the uterine artery of a cow.

Example problem—Womersley solution A pressure gradient is modeled with a simple function,

$$\frac{dP}{dx}(t) = -796 - 1250 \sin(2\pi t) + 531 \cos(4\pi t)$$

Time t is measured in seconds, and the pressure gradient has units of N/m^3. A plot of this function, Fig. 7.7, shows that the period T_0 is 1.0 s and that $\omega = 2\pi$ rad/s.

The blood vessel has the following set of parameters.

Quantity	Symbol	Value	Units
Dynamic viscosity	μ	0.0035	Ns/m^2
Density	ρ	1060	kg/m^3
Kinematic viscosity	ν	3.302×10^{-6}	m^2/s
Vessel radius	R	0.0025	m

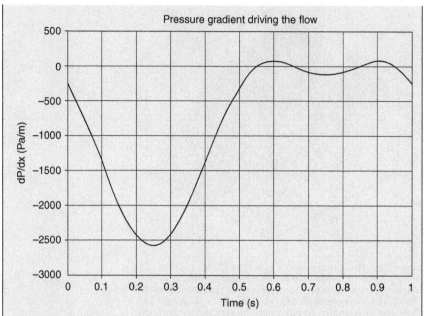

Figure 7.7 Womersley solution example problem: pressure gradient.

Calculate the Womersley solution for pulsatile flow. In addition to the velocity through the pipe cross section at any time, calculate the flow rate through the vessel as a function of time. Present the results with plots.

Solution:

1. We organize the input data. The pressure gradient is already in Fourier series form. It may also be written as

$$\frac{dP}{dx}(t) = \text{Re}(-796 + 1250j\,e^{2\pi jt} + 531e^{4\pi jt})$$

   so that $a_0 = -796$, $a_1 = 1250j$, and $a_2 = 531$.

2. Using the Eq. (1.11) which describes the velocity profile in Poiseuille flow, we write down the steady state (DC analog) solution.

$$u_0 = \frac{1}{4\mu}\,a_0(r^2 - R^2) = -56840\,r^2 + 0.35525$$

3. We compute the parameters needed for Womersley which depend on the vessel geometry and fluid properties. We utilize Eq. (7.48) and write

$$\lambda_n = \sqrt{\frac{j^3\omega n}{\nu}}$$

Pieces of the solution are displayed in the following table.

Term	$n = 1$	$n = 2$
λ_n	$975.4 - 975.4\,j$	$1379 - 1379\,j$
$\dfrac{a_n}{j\rho\omega n}$	0.1877	$-0.03983\,j$
$J_0(\lambda R)$	$-1.0758 + 2.259\,j$	$-5.754 + 0.6115\,j$

4. We calculate and add the pulsatile terms for the velocity solution. Using Eq. (7.56),

$$u_1 = \mathrm{Re}[((-0.03225 - 0.06772j)J_0((975.4 - 975.4j)r)$$
$$- 0.1877)e^{6.283jt}]$$

$$u_2 = \mathrm{Re}[((-0.0007274 + 0.006844j)J_0((1379 - 1379j)r)$$
$$+ 0.03983)e^{12.566jt}]$$

Hence, our solution may be constructed as

$$u(r,t) = u_0 + u_1 + u_2$$

The infinity sign (∞) over the sum in Eq. (7.57) indicates that in the most general case a converging series with an infinity of terms would be needed to represent the arbitrary pressure gradient exactly. In this case, with a simple representation of the pressure gradient using three terms, our solution terminates quickly with just three parts. Two plots illustrate the results. Figure 7.8 shows the three-dimensional portrayal of axial velocity as a function of time and radial position, and Fig. 7.9 shows the velocity versus time for center and midradius.

5. We calculate the flow rate as a function of time by integrating over the cross section, as indicated in Eq. (7.58). As an alternative we could use the formula in Eq. (7.59)

$$Q_n = ((\pi R^2))\,\mathrm{Re}\left[\frac{a_n}{j\omega n\rho}\left\{\frac{2J_1(\lambda R)}{\lambda R\,J_0(\lambda R)} - 1\right\}e^{j\omega nt}\right]$$

To assist a reader who desires to implement this formula, we tabulate the inner term in braces for each harmonic.

n	$\left\{\dfrac{2J_1(\lambda R)}{\lambda R\,J_0(\lambda R)} - 1\right\}$
1	$-0.5783 - 0.3292\,j$
2	$-0.7088 - 0.2467\,j$

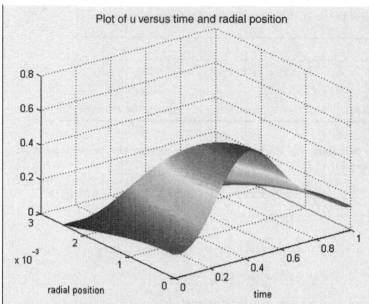

Figure 7.8   Womersley solution of velocity vs. time vs. radial position.

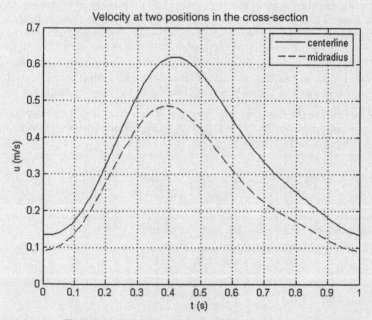

Figure 7.9   Womersley velocity at two points in the cross section vs. time.

As expected, this approach coincides exactly with integrating the velocity over the cross section. Finally, the interested reader may desire to take the approach which Womersley suggests in his paper. For each harmonic,

$$Q_n = \frac{\pi R^4}{\mu} \left(M \frac{M_{10}}{\alpha^2} \right) \sin \left(n\omega t + \phi + \varepsilon_{10} \right)$$

For our two harmonics, we calculate the basic quantities as outlined earlier in this section with the following outcomes.

n	α	M	ϕ (radians)	M_{10} (radians)	ε_{10}
1	3.449	1250	−1.571	0.6655	0.5175
2	4.877	531	−3.142	0.7505	0.3350

Once again, we find that the flow rate curve just calculated agrees with the previous approaches. See Fig. 7.10.

6. We comment on our solution. We note that the velocity lags the pressure gradient. We also note that the velocity peaks at the mid-radius position slightly before the centerline. One important assumption that must be checked is that of laminar flow. We find a peak velocity of 0.62 m/s. We compute

$$\mathrm{Re} = \frac{\rho U_{\max} D}{\mu} = \frac{(1060)(0.62)(0.005)}{0.0033} = 940$$

This is a Reynolds number that definitely lies in the laminar range.

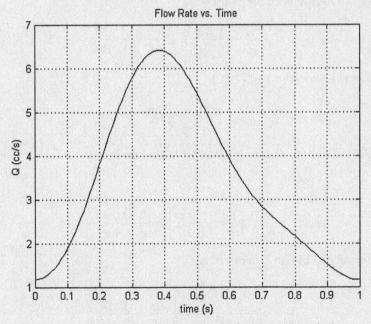

Figure 7.10 Flow rate as a function of time in the Womersley solution.

7.8 Pulsatile Flow in Rigid Tubes—Fry Solution

This solution was published by Greenfield and Fry in 1965 for axisymmetric, uniform, fully developed, horizontal, Newtonian pulsatile flow and is comparable to the Womersley solution in the sense that they are both solutions relating flow rate to pressure gradient for the pulsatile flow condition. With apologies to Dr. Greenfield we will just refer to it as the Fry solution for brevity's sake. Since flow in arteries is pulsatile, this is an important case for human medicine. We will find the Fry solution particularly useful in Chap. 8 in modeling the behavior of an **extravascular catheter/transducer pressure measuring system**. This solution lends itself nicely to the development of an electrical analog to pulsatile flow behavior.

In this solution one may begin from the Navier–Stokes equations where u is velocity in x direction (down the tube centerline), r is radial direction variable (where $r = 0$ on the centerline), v_r is radial velocity, v_Θ is velocity in the transverse direction, and v is μ/ρ or the kinematic viscosity. Recall from Eq. (7.37) the differential equation for axisymmetric, uniform, fully developed horizontal, Newtonian pulsatile flow

$$\frac{a_n e^{j\omega nt}}{\rho} = v\,\frac{\partial^2 u}{\partial r^2} + \frac{v}{r}\frac{\partial u}{\partial r} - \frac{\partial u}{\partial t} = \frac{1}{\rho}\frac{dP}{dx}$$

Next, we divide both sides by v.

$$\frac{\partial^2 u}{\partial r^2} + \frac{1}{r}\frac{\partial u}{\partial r} - \frac{1}{v}\frac{\partial u}{\partial t} = \frac{1}{\mu}\frac{dP}{dx} \tag{7.70}$$

Two of the terms in Eq. (7.70)

$$\frac{\partial^2 u}{\partial r^2} + \frac{1}{r}\frac{\partial u}{\partial r}$$

can be combined, and be written from the chain rule as shown below in Eq. (7.71).

$$\frac{1}{r}\left[(1)\frac{\partial u}{\partial r} + r\frac{\partial^2 u}{\partial r^2}\right] = \frac{1}{r}\frac{\partial\left(r\frac{\partial u}{\partial r}\right)}{\partial r} \tag{7.71}$$

Equation (7.70) can be rewritten:

$$\frac{1}{r}\frac{\partial\left(r\frac{\partial u}{\partial r}\right)}{\partial r} = \frac{1}{v}\frac{\partial u}{\partial t} + \frac{1}{\mu}\frac{dP}{dx} \tag{7.72}$$

If we multiply both sides of Eq. (7.72) by $2\pi r dr$ and integrate from $r = 0$ to R where R is the tube radius we obtain Eq. (7.73).

$$2\pi \int_0^R \frac{\partial\left(r\frac{\partial u}{\partial r}\right)}{\partial r} dr = \frac{1}{v}\int_0^R \left(\frac{\partial u}{\partial t}\right)(2\pi r dr) + \frac{1}{\mu}\frac{dP}{dx}\int_0^R (2\pi r dr) \quad (7.73)$$

The integral of the derivative in the left-hand side of Eq. (7.73) returns the term $r\frac{\partial u}{\partial r}$, so we can rewrite Eq. (7.73) as shown below in Eq. (7.74).

$$2\pi r\frac{\partial u}{\partial r}\bigg|_0^R = \frac{1}{v}\frac{\partial}{\partial t}\int_0^R u(2\pi r dr) + \frac{1}{\mu}\frac{dP}{dx}2\pi\frac{r^2}{2}\bigg|_0^R \quad (7.74)$$

Notice that the integral from zero to R of $u(2\pi r)dr$ is the integral of the velocity multiplied by the differential area so that the term is simply the flow through the tube.

It is possible now to replace the first term on the right-hand side of the equation with the symbol Q, representing flow. After evaluating the last term between 0 and the vessel radius R we arrive at Eq. (7.75).

$$2\pi R\frac{\partial u}{\partial r}\bigg|_R = \frac{1}{v}\frac{\partial Q}{\partial t} + \frac{1}{\mu}\frac{dP}{dx}\pi R^2 \quad (7.75)$$

By dividing both sides of Eq. (7.63) by πR^2 we obtain Eq. (7.76).

$$\frac{2}{R}\frac{\partial u}{\partial r}\bigg|_R = \frac{1}{v\pi R^2}\frac{\partial Q}{\partial t} + \frac{1}{\mu}\frac{dP}{dx} \quad (7.76)$$

Recall that for a Newtonian fluid, the shear stress at the wall is equal to the viscosity times the velocity gradient as shown in Eq. (7.77).

$$\tau_{\text{wall}} = -\mu\frac{\partial u}{\partial r}\bigg|_R \quad (7.77)$$

Substitute Eq. (7.77) into Eq. (7.76) to obtain Eq. (7.78)

$$\frac{2}{R}\left(\frac{-\tau_{\text{wall}}}{\mu}\right) = \frac{1}{v\pi R^2}\frac{\partial Q}{\partial t} + \frac{1}{\mu}\frac{dP}{dx} \quad (7.78)$$

If we rearrange the equation as follows, and multiply it by μ, it will be easier to see what develops.

$$-\frac{dP}{dx} = \frac{\rho}{\pi R^2}\frac{\partial Q}{\partial t} + \left(\frac{2\tau_{\text{wall}}}{R}\right) \quad (7.79)$$

We would like to eliminate the wall stress, τ_{wall}. If we assumed that we had Poiseuille flow, we could use the following expression to substitute for the shear stress at the wall:

$$\tau_{\text{wall}} = \frac{8\mu Q}{2\pi R^3} \tag{7.80}$$

After the replacement of the shear stress term, Eq. (7.79) for the case of Poiseuille flow would be:

$$-\frac{dP}{dx} = \frac{\rho}{\pi R^2}\frac{\partial Q}{\partial t} + \frac{8\mu Q}{\pi R^4} \tag{7.81}$$

Now we have an ordinary, first-order differential equation relating the flow rate and the time rate of change of flow to the pressure gradient. It is one of the simplest differential equations that one encounters. So we would very much like to cast our model into this form, in spite of the fact that we do not actually have Poiseuille flow.

Recall that shear stress in Poiseuille flow depends on the vessel's hydraulic resistance and the flow rate. For the pulsatile flow case, assume that the shear stress depends on both the flow rate and the first derivative of the flow rate as shown by Eq. (7.82).

$$\frac{2\tau_{\text{wall}}}{R} = R_v Q + L_I \frac{dQ}{dt} \tag{7.82}$$

The values for R_v and L_I in Eq. (7.70) are then given by Eqs. (7.83) and (7.84), respectively. Be careful not to confuse the new parameter, named R_v for viscous resistance, with R, radius of the vessel.

$$R_v = R_{\text{viscous}} = C_v \frac{8\mu}{\pi R^4} \tag{7.83}$$

$$L_I = L_{\text{inertia}} = \frac{c_1 \rho}{\pi R^2} \tag{7.84}$$

By substituting Eq. (7.82) into Eq. (7.79) we get:

$$-\frac{dP}{dx} = \frac{\rho}{\pi R^2}\frac{\partial Q}{\partial t} + \frac{8\mu Q}{\pi R^4} \tag{7.85}$$

$$-\frac{dP}{dx} = \frac{\rho}{\pi R^2}\frac{dQ}{dt} + \left(R_v Q + L_I \frac{dQ}{dt}\right) \tag{7.86}$$

Next, substituting the expressions for the viscous resistance term R_v and the inertia term L_I yields the following equation. To further simplify, let $c_u = 1 + c_1$.

$$-\frac{dP}{dx} = \frac{\rho}{\pi R^2}\frac{dQ}{dt} + \frac{c_1\rho}{\pi R^2}\frac{dQ}{dt} + \frac{c_v 8\mu}{\pi R^4}Q \qquad (7.87)$$

$$-\frac{dP}{dx} = c_v\left(\frac{\rho}{\pi R^2}\right)\frac{dQ}{dt} + c_u\left(\frac{8\mu}{\pi R^4}\right)Q \qquad (7.88)$$

Finally, it is possible to simplify the Fry solution to the following first-order ordinary differential equation with terms that represent fluid inertance and fluid resistance as shown below in Eq. (7.89).

$$-\frac{dP}{dx} = L_u\frac{dQ}{dt} + R_v Q \qquad (7.89)$$

where

$$L_u = \frac{(1 + c_1)\rho}{\pi R^2} = \frac{c_u\rho}{\pi R^2} \qquad (7.90)$$

$$R_v = \frac{c_v 8\mu}{\pi\, R^4} \qquad (7.91)$$

The problem with this approach is an acceptable determination of c_u and c_v. This process will be addressed later. For now, we want you to recognize that the Fry solution is very important since it is an analog with a simple electrical circuit, one having an inductor in series with a resistor and a voltage source. This allows very useful modeling analogies to be established. The solutions of these equations will now be addressed.

Differential equations of the kind called for by Fry and Greenfield represent each of the harmonics used in modeling the pressure gradient. They are written as follows:

$$L_u\frac{dQ}{dt} + R_v Q = -\frac{dP}{dx}\bigg|_n = -\mathrm{Re}[a_n e^{j\omega n t}] \qquad (7.92)$$

What we need is the steady state solution to this problem, which takes place after the transient (start-up) portions of the solution have been damped out. As a result, the initial conditions are not needed in this case. It can easily be shown that the steady state solution to the problem looks like

$$Q_n(t) = -\text{Re}[M_n a_n e^{j(\omega nt + \phi_n)}] \tag{7.93}$$

Here the harmonic's **amplification factor** M_n is given by

$$M_n = \frac{1}{\sqrt{(\omega n L_u)^2 + R_v^2}} \tag{7.94}$$

while the phase shift ϕ_n is given by

$$\phi_n = \tan^{-1}\left(-\frac{\omega n L_u}{R_v}\right) \tag{7.95}$$

For a solution using N harmonics, the solution for flow rate will be the Poiseuille flow rate Q_0, coming from the constant term in the pressure gradient added to the sum of the harmonics.

$$Q(t) = Q_0(t) + \sum_{n=1}^{N} Q_n(t) \tag{7.96}$$

The Fry solution requires two coefficients, c_u and c_v. Several options are possible, including matching the solution to experimentally observed flow rates. There is also a formal procedure which causes the Fry solution to match that of Womersley, harmonic by harmonic.

We identify the **phase shift term** in the Womersley solution, ε_{10}, as complementary to ϕ.

Let

$$z = \tan\left(\varepsilon_{10} - \frac{\pi}{2}\right) \tag{7.97}$$

It can be shown that if we stipulate that

$$c_v = \frac{\alpha^2}{8 M_{10} \sqrt{z^2 + 1}} \tag{7.98}$$

and

$$c_u = -\frac{z}{M_{10} \sqrt{z^2 + 1}} \tag{7.99}$$

then the harmonic solution to the Fry equation will coincide with the corresponding Womersley solution. Since M_{10} and ε_{10} are functions of

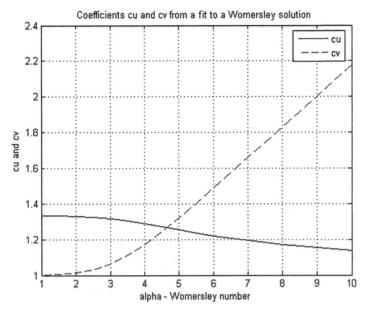

Figure 7.11   The coefficients c_u and c_v used to calculate the Fry parameters.

the number α, we can plot them. See Fig. 7.11. An example problem will now be used to illustrate the construction of a Fry solution which matches the previous Womersley example.

Example problem illustrating the Fry method   This is a continuation of our previous example which uses the pressure gradient,

$$\frac{dP}{dx}(t) = -796 - 1250 \sin(2\pi t) + 531 \cos(4\pi t)$$

Time t is measured in seconds, and the pressure gradient has units of N/m^3. A plot of this function with period $T_0 = 1.0$ s and $\omega = 2\pi$ rad/s is shown in Fig. 7.7. The blood and the artery data are the same as for the previous example.

Quantity	Symbol	Value	Units
Dynamic viscosity	μ	0.0035	Ns/m^2
Density	ρ	1060	kg/m^3
Kinematic viscosity	ν	3.302×10^{-6}	m^2/s
Vessel radius	R	0.0025	m

The problem asks for a Fry solution, giving the flow rate in the artery as a function of time. We will fit the Fry solution to a Womersley solution.

Solution:

1. We organize the input data. From the previous example we know that $a_0 = -796$, $a_1 = -1250\,j$, and $a_2 = 531$.

2. We calculate the steady state flow term with Poiseuille's formula.

$$Q_0 = -\frac{dP}{dx}\bigg|_{n=0}\left(\frac{\pi R^4}{8\mu}\right) = -a_0\left(\frac{\pi R^4}{8\mu}\right) = (796)\left(\frac{3.1416(0.0025)^4}{8(0.0035)}\right)$$

$$= 3.49 \times 10^{-6}\frac{m^3}{s}$$

3. We process each harmonic. We begin with some of the information generated in the previous example problem.

n	a	M_{10}	ε_{10} (radians)
1	3.449	0.6655	0.5175
2	4.877	0.7505	0.3350

It will now be possible to calculate the needed quantities using the following sequence,

$$z = \tan\left(\varepsilon_{10} - \frac{\pi}{2}\right)$$

$$c_v = \frac{\alpha^2}{8M_{10}\sqrt{z^2 + 1}} \quad \text{and} \quad c_u = -\frac{z}{M_{10}\sqrt{z^2 + 1}}$$

$$L_u = c_u\left(\frac{\rho}{\pi R^2}\right) \quad \text{and} \quad R_v = c_v\left(\frac{8\mu}{\pi R^4}\right)$$

$$M_n = \frac{1}{\sqrt{(\omega n L_u)^2 + R_v^2}} \quad \text{and} \quad \phi_n = \tan^{-1}\left(-\frac{\omega n L_u}{R_v}\right)$$

The following table shows the results for the two harmonics used in this problem.

n	c_v	c_u	R_v $\times 10^8$	L_u $\times 10^7$	M_n $\times 10^{-8}$	ϕ_n (radians)
1	1.105	1.306	2.522	7.050	0.1962	-1.0533
2	1.302	1.258	2.972	6.793	0.1106	-1.2358

4. The formulas may now be created and combined. The result is a flow rate identical to the one in the previous example problem.

7.9   Instability in Pulsatile Flow

It is helpful here to point out some useful information concerning stability and **instability** under pulsatile flow conditions. As a rule, flow in large arteries in humans is highly pulsatile. By the time the flow reaches small-diameter arteries and arterioles the pulsatile elements drop out. By the time the flow reaches the capillaries, the flow becomes steady. Flow in large-diameter veins becomes pulsatile once again.

The Reynolds number for flow in humans, even in large-diameter arteries, is generally much less than 2000 and is therefore typically laminar. However, the presence of branching and other fluid wall interactions can sometimes result in local flow instabilities.

When there are disturbances throughout the flow field and throughout the entire oscillatory cycle, this condition constitutes turbulent flow. The presence of vortices at a specific location, or during a specific time in the pulsatile cycle, indicates a local instability only and not turbulent flow. For example, an instability associated with a valvular stenosis is an example of flow that is considered a local flow instability as opposed to turbulent flow.

We end this chapter with a reminder. The models by Womersley and Fry and Greenfield are based on assumptions about the nature of the fluid and the nature of the artery. These assumptions must be checked carefully when looking to apply these techniques to modeling a particular blood flow. Furthermore, the results must be taken for what they are, approximations of a rather crude nature. The best modeling procedures involve a constant interplay between the mathematical models and physical evidence based on experimentation.

Review Problems

1. You are given a pulsatile flow field which is symmetric about the x-axis. The driving force is a pressure gradient,

$$\frac{dP}{dx} = \mathrm{Re}[A e^{j\omega t}]$$

Assume this axis is horizontal. You may assume that the flow is incompressible.

$$u(r,t) = \mathrm{Re}[\,f(r)\, e^{j\omega t}]$$

$$v(r,t) = 0$$

$$w(r,t) = 0$$

What form do the Navier–Stokes equations take?

2. Take the complex number, $z = -50 - 16\,j$. Calculate the following quantities:
   (a) The magnitude of z
   (b) The argument of z in radians and in degrees
   (c) The complex conjugate of z

3. You are given two complex numbers $z_1 = 3 - 6\,j$ and $z_2 = 5 + 2\,j$. Calculate:
   (a) The sum $z_1 + z_2$. Express your answer in both rectangular $(a + bj)$ and polar $(Me^{j\theta})$ forms.
   (b) The difference $z_2 - z_1$. Express your answer in both forms.
   (c) The product $z_1 z_2$. Express your answer in both forms.
   (d) The quotient z_1/z_2. Express your answer in both forms.

   (The use of a computer is strongly suggested for the remaining problems.)

4. Bessel functions of the kind discussed in this chapter can take complex numbers as arguments. Say you are given a number $z_0 = 50 - 20\,j$. Using your favorite software (Matlab is suggested), calculate:
   (a) $J_0(z_0)$
   (b) The magnitude of the above number
   (c) The argument of the above number

5. Take a look at the following triangular pulse function. See Fig. 7.12. Its mathematical specification over one period, $T_0 = 1.6$, is

$$f(t) = \begin{cases} 2.5t & t \le 0.4 \\ 2 - 2.5t & 0.4 < t \le 0.8 \\ 0 & \text{otherwise} \end{cases}$$

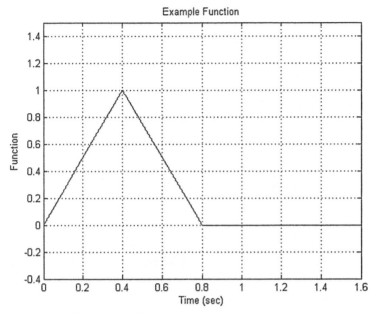

Figure 7.12 Waveform. Prob. 5.

Calculate the trigonometric Fourier series. Give at least four terms.

6. Here is a sinusoidal pulse function. See Fig. 7.13. Its mathematical specification over one period, $T_0 = 1.2$, is

$$f(t) = \left\{ \begin{array}{ll} \sin(2\pi t) & t \leq 0.5 \\ \\ 0 & \text{otherwise} \end{array} \right\}$$

The coefficients for the complex series,

$$f(t) = \sum_{n=-\infty}^{N=\infty} a_n e^{jn\omega t}$$

are defined by

$$a_n = \frac{1}{T_0} \int_0^{T_0} f(t) e^{-jn\omega t} dt$$

Calculate the complex coefficient Fourier series and the trigonometric Fourier series. Observe the relationships between the coefficients in the two series.

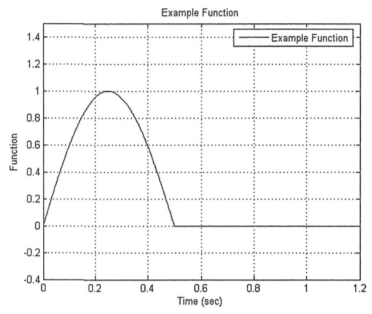

Figure 7.13   Prob. 6.

7. A simple illustration of a two-term solution to the pulsatile flow problem is derived from the following pressure gradient:

$$\frac{dP}{dx}(t) = -500 + \mathrm{Re}(500e^{7.854jt})$$

Time t is measured in seconds, and the pressure gradient has units of N/m^3. A plot of this function (Fig. 7.14) shows that the period T_0 is 0.8 s and that $\omega = 2.5\pi$ rad/s.

The blood vessel has the following set of parameters.

Quantity	Symbol	Value	Units
Dynamic viscosity	μ	0.0033	Ns/m^2
Density	ρ	1060	kg/m^3
Kinematic viscosity	ν	3.113×10^{-6}	m^2/s
Vessel radius	R	0.0020	m

Calculate the Womersley solution for pulsatile flow. In addition to the velocity through the pipe cross section at any time, calculate the flow rate through the vessel as a function of time.

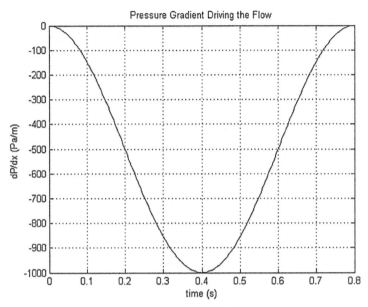

Figure 7.14   Pressure gradient for Prob. 7.

8. Using a computer program, produce the actual plot of the Womersley solution in the example problem in Sec. 7.7.

9. In his historic 1954 paper, Womersley presents the modeling of the pulsatile flow in the femoral artery of a dog. The pressure gradient is described by the following table of Fourier coefficients, along with the magnitudes and phase angles for each harmonic. In the case of the latter, we have converted the angles in degrees given by Womersley to radians.

N	Cosine coefficient	Sine coefficient	M_n	ϕ_n(radians)
1	0.8781	−0.7432	1.1050	0.7024
2	0.5415	1.4327	1.5316	−1.2094
3	−0.7946	0.5508	0.9668	0.6062
4	−0.2375	−0.1588	0.2857	−0.5896
5	0.0125	−0.2818	0.2821	1.5222
6	−0.1917	−0.0167	0.1924	−0.0867

The coefficients given correspond to measurements in mmHg. For consistent units and flow in the correct direction they must be multiplied by −133.32. The period is $T_0 = 1/3$ s. The vessel radius is 0.0015 m, the viscosity is 0.04 P, and the density is 1.05 g/mL. Using a method of your choice, calculate and plot the flow rate as a function of time over one period. You may wish to compare your results with the plots on page 559 of Womersley's paper. (Your plotting software is probably much better than Womersley's, so don't expect perfect agreement.) If you wish to avoid the use of Bessel functions, you can employ the table below which lists the quantities α, M_{10}, and ε_{10}. The latter angles are presented in radians.

n	α	M'_{10}	ε_{10}
1	3.34	0.6651	1.0815
2	4.72	0.7436	0.6964
3	5.78	0.7839	0.5521
4	6.67	0.8096	0.4712
5	7.46	0.8278	0.4171
6	8.17	0.8416	0.3752

10. A previous problem specifies pressure gradient:

$$\frac{dP}{dx}(t) = -500 + \mathrm{Re}(500e^{7.854jt}) = -500 + 500 \cos{(7.854t)}$$

where time t is measured in seconds, and the pressure gradient has units of N/m^3. The blood vessel has the following set of parameters:

Quantity	Symbol	Value	Units
Dynamic viscosity	μ	0.0033	Ns/m^2
Density	ρ	1060	kg/m^3
Kinematic viscosity	ν	3.113×10^{-6}	m^2/s
Vessel radius	R	0.0020	m

Calculate the Fry solution to the problem. Use the Womersley solution to calibrate your Fry solution. Present the results with plots.

11. The pressure gradient in the artery of a pig has been measured and digitized. The sampling rate was 512/s. Figure 7.15 is a plot of that data. The data are accessible at the text website at http://www.mhprofessional.com/product.php?isbn=0071472177. Represent theses data with a 10-term Fourier series. Use the Fast Fourier Transform to calculate the Fourier coefficients.

12. Given the vessel data from prob. 10 and pressure gradient data from prob. 11, construct a 10-term Womersley solution to the previous problem.

13. Continue the previous problem by constructing a Fry solution which matches the Womersley solution.

14. Using a software tool of your choice, create a computer program which inputs the Fourier coefficients and fundamental frequency of the pressure

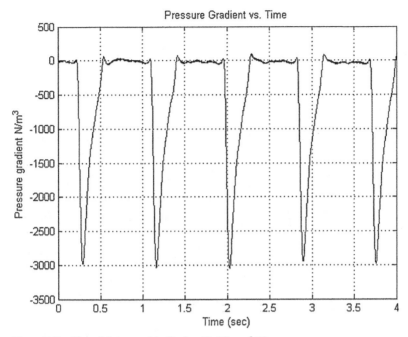

Figure 7.15   Plot of data used in Probs. 11, 12, and 13.

gradient, the radius of the vessel, and the viscosity and density of the blood. The program should then output a plot of the flow rate over one period.

Bibliography

Greenfield, JC, Fry, DL, "Relationship between Instantaneous Aortic Flow and the Pressure Gradient," *Circ. Res.*, 17, Oct 1965.

Munson, BR, Young, DF, Okiishi, T, *Fundamentals of Fluid Mechanics*, John Wiley & Sons, New York, 1994.

Rainville, ED, Bedient, PE, *Elementary Differential Equations*, 5th ed., Macmillan, New York, 1974.

Del Toro, V, *Principles of Electrical Engineering*, 2nd ed., Prentice-Hall, Englewood Cliffs, NJ, 1972.

Womersley, JR, "Method for the Calculation of Velocity, Rate of Flow and Viscous Drag in Arteries When the Pressure Gradient Is Known," *J. Physiol.*, 127, 553−563, 1955.

Flow and Pressure Measurement

8.1 Introduction

From the measurement system for measuring **hypertension** in humans with an indirect method, to the system for measurement of flow and pressure waveforms in the uterine artery of cows, flow and pressure measurement systems have wide-ranging applications in biology and medicine. A few of the varied methods for making these measurements are discussed below.

Measurements of flow and pressure are two of the most important measurements in biological and medical applications. Because there is a complex control system in the human body controlling pressure and flow, the interactions of the measuring system with the physiological control system increase the complexity of the measurement. The invasive nature of flow and pressure measurement also makes it difficult to obtain accurate measurements in some cases. Sometimes, a noninvasive or **indirect measurement** is possible.

According to the American Heart Association Council on High Blood Pressure Research, accurate measurement of blood pressure is critical to the diagnosis and management of individuals with hypertension. Dr. Thomas Pickering et al. (2005) write, "The ausculatory (indirect measurement) technique with a trained observer and mercury sphygmomanometer continues to be the method of choice for measurement in the office."

8.2 Indirect Pressure Measurements

An important example of a noninvasive, indirect measurement is the measurement of blood pressure using a sphygmomanometer. This type of measurement is routinely used in clinical practice because of its relative ease and low cost.

A **sphygmomanometer** has a cuff that can be wrapped around a patient's arm and inflated with air. The device includes a means for pressurizing the cuff, and a gauge for monitoring the pressure inside the cuff. When the pressure inside the cuff is increased above the patient's systolic pressure, the cuff squeezes the arm, and blood flow through the vessels directly under the cuff is completely blocked. The pressure inside the cuff is released slowly, and when the cuff pressure falls below the patient's systolic pressure, blood can spurt through arteries in the arm beneath the cuff. The operator measuring blood pressure typically places a stethoscope distal to the cuff and just over the brachial artery. Turbulent flow of blood through the compressed artery makes audible sounds, known as **Korotkoff sounds**. At the time these Korotkoff sounds are initially detected, the blood pressure is noted, and this value becomes the measurement of the systolic pressure.

The nature of the sound heard through the stethoscope changes as the sphygmomanometer pressure continues to drop. The sounds change from a repetitive tapping sound to a muffled rumble and then the sound finally disappears. The pressure is once again noted at the instant that the sounds completely disappear. This value is the diastolic blood pressure.

Indirect measurement of blood pressure using a sphygmomanometer has the advantage of being noninvasive, inexpensive, and reliable. One disadvantage of the system is the relative inaccuracy of the system associated with operator error, relatively low resolution, and the difficulty of measuring the diastolic pressure. Since the measurement of diastolic pressure depends on the detection of the absence of sound, the method is inherently dependent on the ability of the operator to clearly hear very quiet sounds and to judge the point in time at which those sounds disappear.

8.2.1 Indirect pressure gradient measurements using Doppler ultrasound

By using **echocardiography,** it is possible to make an indirect measurement of pressure gradients at specific locations in the cardiovascular system that are estimated from changing blood velocities. Özyurda et al. (2005) used echocardiography to indirectly measure pressure gradients in 400 patients with aortic or mitral valve replacement with prosthetic heart valves. The Bernoulli equation (see also Sec. 6.4.3) was used to calculate pressure gradient across the prosthesis.

$$\Delta P = \frac{\rho}{2}\left(V_2^2 - V_1^2\right)$$

where ΔP was the pressure gradient, or pressure drop across the valve, V_2 represents blood velocity through the valve, V_1 represents the left ventricular outflow tract velocity (the velocity leaving the ventricle), and ρ is blood density.

The authors of the study calculated pressure gradients in mmHg using multiple instantaneous velocities through the valve, throughout the velocity profile. The average of all such instantaneous pressure gradients was reported as the mean pressure gradient. While the application of the Bernoulli equation ignores frictional losses across the valve, this method provides a noninvasive, clinically meaningful estimate of pressure gradient across the valve. The authors of the study reported the equation they used as the "modified Bernoulli equation," which was reported as:

$$\Delta P = \frac{\rho}{2}\left(V_2^2 - V_1^2\right) = \frac{1060 \text{ kg/m}^3}{2(133)\text{Pa/mmHg}}\left(V_2^2 - V_1^2\right)\frac{\text{m}}{\text{s}} = 4\left(V_2^2 - V_1^2\right)$$

where velocities are reported in m/s and pressure gradients are reported in mmHg.

8.3   Direct Pressure Measurement

When more accurate pressure measurements are required, it becomes necessary to make a direct measurement of blood pressure. Two methods that are described below involve using either an intravascular catheter with a strain gauge measuring transducer on its tip or an extravascular transducer connected to the patient by a saline-filled tube.

8.3.1   Intravascular—strain gauge tipped pressure transducer

Transducers are devices that convert energy from one form to another. It is often desirable to have electrical output represent a parameter like blood pressure or blood flow. Typically, these devices convert some type of mechanical energy to electrical energy. For example, most pressure-measuring transducers convert stored energy in the form of pressure into electrical energy, perhaps measured as a voltage.

A pressure transducer very often uses a strain gauge to measure pressure. We can begin the topic of strain gauge-tipped pressure transducers by considering strain. Figure 8.1 shows a member under a simple uniaxial load in order to demonstrate the concept of strain. As the load causes the member to stretch some amount ΔL, strain can be measured as $\Delta L/L$. A strain gauge could be bonded onto a diaphragm on a catheter tip.

Two advantages of a strain gauge pressure transducer, compared to an extravascular pressure transducer described in Sec. 8.3.2, are good frequency response, and fewer problems associated with blood clots.

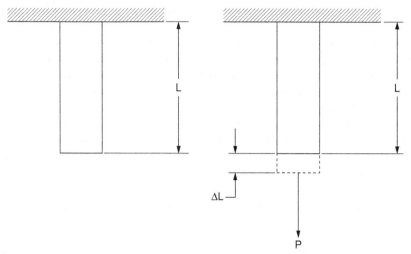

Figure 8.1   Demonstration of strain in a tensile specimen, $\Delta L/L$.

Some disadvantages might include the difficulty in sterilizing the transducers, the relatively higher expense, and transducer fragility.

As the strain gauge lengthens due to the load, the diameter of the wire in the strain gauge also decreases and therefore the resistance of the wire changes. Figure 8.2 shows a drawing of a strain gauge with a load F applied. It is possible to calculate the resistance of a wire based on its length, its cross-sectional area, and a material property known as resistivity. Equation (8.1) relates strain gauge resistance R to those properties.

$$R = \rho L/A \qquad (8.1)$$

where R = resistance, in Ohms, Ω
ρ = resistivity, in Ohm · meters, $\Omega \cdot m$

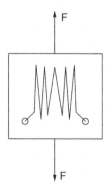

Figure 8.2 Strain gauge with force F applied.

L = length in meters, m

A = wire cross-sectional area in square meters, m^2

We can now examine the small changes in resistance due to the changes in length and cross-sectional area by using the chain-rule to take the derivative of both sides of Eq. (8.1). The result is shown in Eq. (8.2).

$$dR = \frac{\rho}{A}dL - \frac{\rho L}{A^2}dA + \frac{L}{A}d\rho \qquad (8.2)$$

We can now divide Eq. (8.2) by $\rho L/A$ to get the more convenient form of the equation shown in Eq. (8.3).

$$\frac{dR}{R} = \frac{dL}{L} - \frac{dA}{A} + \frac{d\rho}{\rho} \qquad (8.3)$$

If you stretch a wire using an axial load, the diameter of the wire will also change. As the wire becomes longer, the diameter of the wire becomes smaller. Poisson's ratio is a material property which relates the change in diameter to the change in length of wire for a given material. Poisson's ratio is typically written as the character ν, which students should not confuse with kinematic viscosity, which is also typically written as ν. Poisson's ratio is defined in Eq. (8.4).

$$\text{Poisson's ratio} = \nu = \frac{-dD/D}{dL/L} \qquad (8.4)$$

In Eq. (8.4), D is the strain gauge wire diameter, and L is the strain gauge wire length.

It is also useful here to show that, since $A = \frac{\pi}{4}D^2$, dA/A is related to dD/D so that we can write Eq. (8.3) in the more convenient terms of wire diameter instead of cross-sectional area.

$$\frac{dA}{A} = \frac{\frac{\pi}{4}(D_2^2 - D_1^2)}{\frac{\pi}{4}D^2} = \frac{\frac{\pi}{4}(D_2 - D_1)(D_2 + D_1)}{\frac{\pi}{4}D^2} = \frac{\frac{\pi}{4}(D_2 - D_1)(2D)}{\frac{\pi}{4}D^2} \qquad (8.5)$$

In Eq. (8.5), dD is equal to $D_2 - D_1$ and $D_2 + D_1 = 2D$, where D is the average diameter between D_1 and D_2. Therefore, the relationship between dA and dD is shown by Eq. (8.6).

$$\frac{dA}{A} = \frac{2dD}{D} \qquad (8.6)$$

Now it is possible to write an equation for the change in resistance divided by the nominal resistance as shown in Eq. (8.7).

$$\frac{dR}{R} = \frac{dL}{L} - \frac{2dD}{D} + \frac{d\rho}{\rho} \tag{8.7}$$

or by combining with the definition of Poisson's ratio $v = \dfrac{-dD/D}{dL/L}$

$$\frac{dR}{R} = \frac{dL}{L} + 2v\frac{dL}{L} + \frac{d\rho}{\rho} = (1 + 2v)\frac{dL}{L} + \frac{d\rho}{\rho} \tag{8.8}$$

The term $(1 + 2v)dL/L$ represents a change in resistance associated with dimensional change of the strain gauge wire. The second term $d\rho/\rho$ is the term that represents a change in resistance associated with piezoresistive effects, or change in crystal lattice structure within the material of the wire.

The gauge factor for a specific strain gauge is defined by Eq. (8.9).

$$\text{Gauge factor} \equiv G = \left(\frac{\Delta R}{R} \bigg/ \frac{\Delta L}{L}\right) = (1 + 2v) + \left(\frac{\rho}{\Delta\rho} \bigg/ \frac{\Delta L}{L}\right) \tag{8.9}$$

For metals like nichrome, constantan, platinum-iridium, and nickel-copper the term $(1 + 2v)$ dominates the gauge factor. For semiconductor materials like silicon and germanium, the second term

$$\left(\frac{\rho}{\Delta\rho} \bigg/ \frac{\Delta L}{L}\right)$$

dominates the gauge factor.

A typical gauge factor for nichrome wire is approximately 2, and for platinum-iridium approximately 5.1. Strain gauges using semiconductor materials have a gauge factor that is two orders of magnitude greater than that of metal strain gauges. Semiconductor strain gauges are therefore more sensitive than metallic strain gauges.

If we know the gauge factor of a strain gauge, and if we can measure $\Delta R/R$, then it is possible to calculate the strain $\Delta L/L$. However, the change in resistance ΔR is a very small change. Even though it is an electrical measurement, we need a strategy to detect such a small change in resistance.

A Wheatstone bridge is a device that is designed to measure very tiny changes in resistance. See the circuit diagram in Fig. 8.3. In Fig. 8.3, I_1 is the current flowing through the two resistors, R_1. The current

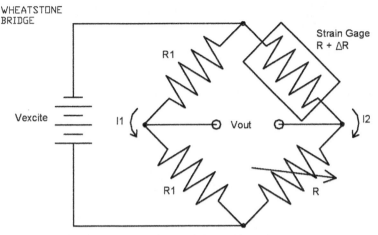

Figure 8.3   A bridge for measuring small changes in resistance.

flowing through the two resistors can be shown to be the excitation voltage driving the circuit divided by $2R_1$ as shown in Eq. (8.10).

$$I_1 = V_e/(2 \cdot R_1) \tag{8.10}$$

The voltage at point 1, between the two resistors is:

$$E_1 = I_1 \cdot R_1 = V_e/2 \tag{8.11}$$

The current I_2, flowing through the right side of the circuit, through the strain gauge and the potentiometer is:

$$I_2 = V_e/(2R + \Delta R) \tag{8.12}$$

The voltage at point two, on the right side of the circuit, between the strain gauge and potentiometer is:

$$E_2 = I_2 \cdot R = V_e \frac{R}{2R + \Delta R} \tag{8.13}$$

V_{out} for the bridge is now given by the difference between the voltages at points 1 and 2. See Eqs. (8.14) through (8.16).

$$E_1 - E_2 = (V_e/2)(1 - (2R/(2R + \Delta R))) \tag{8.14}$$

$$E_1 - E_2 = (V_e/2)(\Delta R/(2R + \Delta R)) \tag{8.15}$$

and since $2R$ is much, much greater than ΔR

$$E_1 - E_2 = (V_e/4)(\Delta R/R) \tag{8.16}$$

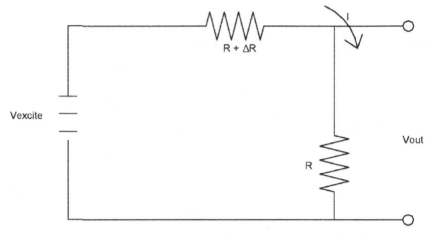

Figure 8.4 A voltage measuring circuit that does not use a bridge. This circuit cannot accurately measure small changes in resistance.

Therefore, to measure ($\Delta R/R$) we can use a bridge and measure ($V_{out}/V_{excitation}$). It is possible to obtain $\Delta R/R$ by multiplication of ($V_{out}/V_{excitation}$) by 4. Now, if we divide $\Delta R/R$ by the gauge factor G, we will obtain the strain.

If you question the necessity of the bridge in our measurement, compare the circuit in Fig. 8.4. If you use a known resistance in series with the strain gauge and measure resistance across the strain gauge, by measuring voltage, you will obtain the following. The current I flowing through the circuit is:

$$I = V_e/(2R + \Delta R) \qquad (8.17)$$

The output voltage of the circuit that is used to measure the change in resistance is:

$$V_{out} = IR = (V_e \cdot R)/(2R + \Delta R) \qquad (8.18)$$

Since $R/2R$ is $\gg R/\Delta R$, this circuit yields a fairly constant output of $^1/_2 V$.

Finally, strain is proportional to the strain measured by the transducer and the strain gauge can be calibrated to output a voltage proportional to pressure. One can imagine a pressure transducer with a diaphragm that moves depending on the pressure of the fluid inside the transducer. A strain gauge mounted on the diaphragm of the transducer measures the displacement, which is proportional to pressure.

In a strain gauge-tipped pressure transducer, a very small strain gauge is mounted on the tip of a transducer, and the strain gauge tip deflects proportionally to the pressure measured by the strain gauge. The resultant strain may also be converted to a voltage output, which may be calibrated to the pressure being measured.

8.3.2 Extravascular—catheter-transducer measuring system

Figure 8.5 shows a schematic of an **extravascular pressure transducer.** The transducer is connected to a long, thin tube called a catheter. The catheter can transmit pressure from the blood vessel of interest to the extravascular pressure transducer. The pressure is transmitted through a column of heparinized saline. Heparin is used to prevent the blood from clotting and clogging the end of the catheter.

Although the extravascular pressure measurement system is a nice compromise between relatively low cost and accuracy, there are several potential problems associated with the system that will be mentioned here.

- Air bubbles in the system, when present, have an important effect on the system's ability to measure high-frequency components of the pressure waveform. When you have air bubbles in the brake lines of your automobile, the brakes feel spongy and are less responsive. In the same fashion, an extravascular transducer with air bubbles in the catheter will be spongy and may not respond accurately to the pressure.

- Blood clots can form in the catheter. The blood clots will restrict flow and either plug the catheter or cause a significant pressure drop between the vessel and the extravascular transducer.

- The use of the extravascular pressure transducers requires a surgical procedure. This disadvantage is also present in other direct pressure measurements but is not a disadvantage for indirect pressure measurements.

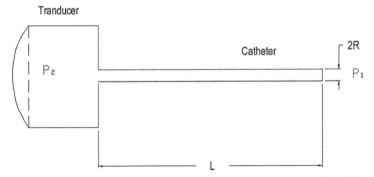

Figure 8.5 An extravascular pressure transducer connected to a long thin catheter which can be inserted into a blood vessel to transmit the pressure in that vessel.

8.3.3 Electrical analog of the catheter measuring system

In Chap. 7 Sec. 7.8, a solution was developed that was published by Greenfield and Fry in 1965 that shows the relationship between flow and pressure for axisymmetric, uniform, fully developed, horizontal, Newtonian pulsatile flow. The Fry solution is particularly useful when considering the characteristics of a transducer and catheter measuring system. By developing an electrical analog to a typical pressure measuring catheter, it will be possible for us to use some typical, well-known solutions to RLC circuits to characterize things like the natural frequency and dimensionless damping ratio of the system. From our circuit analog, we will be better able to understand the limitations of our pressure measuring system and to predict important characteristics.

It was possible to simplify the Fry solution to the following first-order ordinary differential equation with terms that represent fluid inertance and fluid resistance as was shown below and in Chap. 7, Eq. (7.89).

$$\frac{P_1 - P_2}{\ell} = L\frac{dQ}{dt} + R_v Q \tag{7.89}$$

In Eq. (7.89), P_1 represents the pressure in the artery being measured, P_2 represents the pressure at the transducer, ℓ represents the length of the catheter, Q is the flow rate of the saline in the catheter, and dQ/dt is the time rate of change of the flow rate. Hydraulic inertance and hydraulic resistance are represented by L and R_v, respectively, and are defined in Eqs. (8.19) and (8.20).

$$L = \frac{(1 + c_1)\rho}{\pi R^2} = \frac{c_u \rho}{\pi R^2} \tag{8.19}$$

$$R_V = \frac{c_v 8\mu}{\pi R^4} \tag{8.20}$$

In Eq. (8.19) ρ represents the fluid (saline) density in the catheter and R represents the radius of the catheter. Three empirical proportionality constants are represented by c_u, c_1, and c_v. In Eq. (8.20) μ represents fluid viscosity (saline viscosity).

The volume compliance of the transducer C represents the stiffness of the transducer or the change in volume inside the transducer corresponding to a given pressure change. The volume compliance is written in Eq. (8.21).

$$C = \frac{dV}{dP_2} \tag{8.21}$$

By separating variables and integrating with respect to time, it is possible to solve for flow rate Q as a function of the change in the pressure in the transducer as shown in Eqs. (8.22) and (8.23).

$$dV = C\, dP_2 \tag{8.22}$$

$$Q = \frac{dV}{dt} = C\frac{dP_2}{dt} \tag{8.23}$$

The time rate of change of Q, as a function of compliance and P_2, can now be written:

$$\frac{dQ}{dt} = C\frac{d^2P_2}{dt^2} \tag{8.24}$$

Now it becomes possible to substitute Eqs. (8.23) and (8.24) into Eq. (7.89) above.

$$LC\frac{d^2P_2}{dt^2} + R_vC\frac{dP_2}{dt} = \frac{P_1 - P_2}{\ell} \tag{8.25}$$

L and R_v are defined by Eqs. (8.19) and (8.20), respectively. C is the volume compliance of the transducer. The length of the catheter is represented by ℓ, and the pressures in the blood vessel and the transducer are P_1 and P_2, respectively.

Next rewrite the derivative terms in the somewhat simplified form where $\frac{dP}{dt}$ is written as $\dot{P}$ and $\frac{d^2P}{dt^2} = \ddot{P}$ and we arrive at Eq. (8.26).

$$\ell LC\ddot{P}_2 + \ell R_vC\dot{P}_2 + P_2 = P_1 \tag{8.26}$$

Now we can define a term, E, that is equal to $1/C$ or the inverse of the volume compliance of the transducer. The term E is known as the volume modulus of elasticity of the transducer. Then, by multiplying both sides of the equation by the catheter length, ℓ, we arrive at Eq. (8.27). Notice that Eq. (8.27) is a second-order, linear, ordinary, differential equation with driving function EP_1. This type of equation is typical in many, many types of electrical and mechanical applications and one that we will use several times.

$$\underbrace{\ell L\ddot{P}_2}_{m} + \underbrace{\ell R_v\dot{P}_2}_{b} + \underbrace{E}_{k} P_2 = EP_1 \tag{8.27}$$

For the mechanical analog system shown in Fig. 8.6, the equation defines a typical spring, mass, damper system where the spring constant is

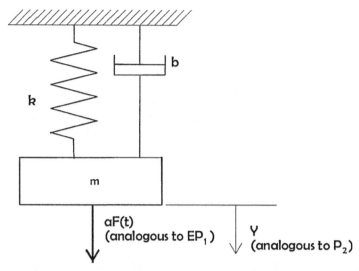

Figure 8.6   A spring, mass, damper analog to the extravascular pressure measuring system. The driving force for the system in the picture is $aF(t)$.

$k = E$, the damping coefficient for the damper is $b = \ell R_v$, and the mass term is $m = \ell L$. The canonical form of the second-order differential equation describing the spring, mass, damper system, which practically all engineers have seen, is shown in Eq. (8.28).

$$m\ddot{y} + b\dot{y} + ky = aF(t) \tag{8.28}$$

8.3.4   Characteristics for an extravascular pressure measuring system

For all second-order systems, there are several system characteristics that may be of interest. In this section, we discuss those characteristics, including static sensitivity, undamped natural frequency, and damping ratio.

For the mechanical system from Eq. (8.28), the characteristics are well known and are written in Eqs. (8.29), (8.30), and (8.31).

In Eq. (8.29), y represents displacement of the mass shown in the mechanical spring mass and damper system in Fig. 8.6. The figure shows a time-varying driving force $aF(t)$ driving the mass up and down while it is attached to a spring with spring rate k, and a damper with associated constant b.

$$\text{static sensitivity} \equiv k = a/y \tag{8.29}$$

$$\text{undamped natural frequency} \equiv \omega_n = \sqrt{\frac{k}{m}} \frac{\text{rad}}{\text{s}} \tag{8.30}$$

$$\text{damping ratio} \equiv \zeta = \frac{b}{2\sqrt{km}} = \frac{b}{b_{cr}} \qquad (8.31)$$

By making the appropriate substitutions from Eq. (8.27), we can now solve for the specific characteristics associated with our extravascular pressure measurement system.

$$\underbrace{\ell L}_{m} \ddot{P}_2 + \underbrace{\ell R_v}_{b} \dot{P}_2 + \underbrace{E}_{k} P_2 = \underbrace{E}_{a} P_1 \qquad (8.27)$$

For the static system, displacement y does not change. In the pressure measuring system, the pressure does not change. All of the higher-order terms like $\dot{y}$, $\ddot{y}$, $\dot{P}_2$ and $\ddot{P}_2$ are now zero. For the pressure measuring system, it is desirable that for every pressure input an equivalent pressure output occurs, so let us design the static gain to be unity. Equation (8.28) for the static system is:

$$\underbrace{m\ddot{y}}_{0} + \underbrace{b\dot{y}}_{0} + ky = aF(t) \qquad \text{or} \qquad ky = aF(t) \qquad (8.28)$$

For the analogous pressure measuring system $k = E$ and $a = E$ so the static sensitivity is one, or E/E.

Again, this means simply that the measurement system is designed to measure the input value and repeat it as the output value.

The undamped natural frequency and dimensionless damping ratio of the system are given by Eqs. (8.32) and (8.33).

$$\text{Undamped natural frequency} \equiv \omega_n = \sqrt{\frac{E}{\ell L}} \frac{\text{rad}}{\text{s}} = \sqrt{\frac{E\pi R^2}{\ell c_u \rho}} \frac{\text{rad}}{\text{s}} \qquad (8.32)$$

$$\text{Damping ratio} \equiv \zeta = \frac{\ell R_v}{2\sqrt{E\ell L}} = \frac{\ell \dfrac{c_v 8\mu}{\pi R^4}}{2\sqrt{E\ell \dfrac{c_u \rho}{\pi R^2}}} \qquad (8.33)$$

Sections 8.3.6 and 8.3.7 describe two examples of second-order, pressure measuring systems. Case 1 is an undamped catheter measuring system, and Case 2 is the undriven, damped system.

8.3.5 Example problem—characteristics of an extravascular measuring system

A liquid-filled catheter, 100 cm long, with an internal radius of 0.92 mm, is connected to a strain gauge pressure sensor. The curve in Fig. 8.7 shows a

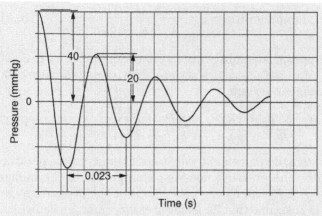

Figure 8.7 Pop test from an extravascular pressure measuring catheter.

pop test of the system. (The pop test is further discussed in Sec. 8.3.8). (a) Find the damped natural frequency, undamped natural frequency and the damping ratio. (b) Use the values found in part (a) of this problem to estimate the modulus of elasticity of the pressure sensor diaphragm. Use $c_u = 1.1$ and density of the fluid $= 1000 \text{ kg/m}^3$.

Solution The period of the signal from the pop test $T = 0.023$ s. Therefore, the damped natural frequency is $1/T = 1/0.023$ s $= 43.5$ rad/s or 415 Hz. The logarithmic decrement from the graph is

$$\delta = \ln\left[\frac{P_2(0)}{P_2(1)}\right] = \ln\left[\frac{40}{20}\right] = 0.693$$

From Eq. (8.54), the **damping ratio** is

$$\zeta = \sqrt{\left[\frac{\delta^2}{4(\pi)^2 + \delta^2}\right]} = \sqrt{\left[\frac{(0.693)^2}{4(\pi)^2 + (0.693)^2}\right]} = 0.110$$

The **damped natural frequency** is

$$\frac{1}{\text{period}} = \frac{1}{0.023} = 43.5 \text{ Hz} \quad \text{or} \quad 43.5(2\pi) = 273 \frac{\text{rad}}{\text{s}}$$

The **natural frequency** can now be found from Eq. (8.55)

$$\omega_n = \frac{\omega_D}{\sqrt{1 - \zeta^2}} = \frac{43.5 \text{ Hz}}{\sqrt{1 - 0.11^2}} = 43.7 \text{ Hz or } 275 \frac{\text{rad}}{\text{s}}$$

8.3.6   Case 1: the undamped catheter measurement system

For the special case of a driven, undamped catheter measuring system we can now estimate some of the characteristics of the system. An undamped system would be one in which the viscosity of the fluid in the catheter, and therefore the corresponding hydraulic resistance in the catheter, are very small; therefore close to zero. A practical example of an undamped measuring system is one in which the viscosity is very small compared to the mass of the system. A relatively large diameter catheter, with no air bubbles and with a relatively low viscosity fluid like water, and relatively low total volume displacement would behave in this way. A system like this could have a very fast response time but would be susceptible to "catheter whip," or noise associated with accelerations at the catheter tip.

For this case recognize that viscosity μ, viscous resistance R_v, and the damping coefficient b are all approximately equal to zero. Then recall the expression for undamped natural frequency in Eq. (8.32). We can already see that the damping ratio is zero for this undamped system, and we could predict the natural frequency from Eq. (8.32).

$$\mu \cong 0 \qquad R_v \cong 0 \qquad b \cong 0$$

$$\text{undamped natural frequency} \equiv \omega_n = \sqrt{\frac{E}{\ell L}}\,\frac{\text{rad}}{\text{s}} = \sqrt{\frac{E\pi R^2}{\ell c_\mu \rho}}\,\frac{\text{rad}}{\text{s}} \tag{8.32}$$

$$\text{damping ratio} \equiv \zeta = \frac{\ell R_v}{2\sqrt{E\ell L}} = 0 \tag{8.33}$$

Going a step further, we would also like to predict the output of the pressure measuring system for every input. If P_1 is the input pressure to the system, then we would like the pressure output P_2 to be equal to P_1 for any input frequency. Let's set the input, or driving pressure, to be $P_1 \cos(\omega t)$. The output of the system or the pressure at the transducer is now $P_2 \cos(\omega t)$. Recall the governing differential equation from Eq. (8.27).

$$\ell L \ddot{P}_2 + \ell R_v \dot{P}_2 + E P_2 = E P_1 \tag{8.27}$$

Now after solving for the first and second derivative of P_2, the expressions below can be substituted into Eq. (8.27).

$$P_2 = P_2 \cos(\omega t) \tag{8.34}$$

$$\dot{P}_2 = -\omega P_2 \sin(\omega t) \tag{8.35}$$

$$\ddot{P}_2 = -\omega^2 P_2 \cos(\omega t) \tag{8.36}$$

Equation (8.27) now becomes Eq. (8.37). Note that P_2 now represents a constant.

$$\ell L(-\omega^2)P_2\cos(\omega t) + \ell R_v(-\omega)P_2\sin(\omega t) + EP_2\cos(\omega t)$$
$$= EP_1\cos(\omega t) \tag{8.37}$$

Since R_v is equal to zero for this case, the second term drops out and we can divide through by $\cos(\omega t)$. The result is Eq. (8.38).

$$\ell L(-\omega^2)P_2 + EP_2 = EP_1 \tag{8.38}$$

Solving for P_2 yields Eq. (8.39).

$$P_2 = \frac{EP_1}{(E - \ell L\omega^2)} = \frac{P_1}{\left(1 - \dfrac{\ell L\omega^2}{E}\right)} \tag{8.39}$$

Now using the definition of ω_n from Eq. (8.32) we can finally solve for the transfer function, P_2/P_1 as shown in Eq. (8.41).

$$\frac{1}{\omega_n^2} = \frac{\ell L}{E} \tag{8.40}$$

$$\frac{P_2}{P_1} = \frac{1}{\left[1 - \dfrac{\omega^2}{\omega_n^2}\right]} \tag{8.41}$$

Finally, Fig. 8.8 shows the response of the second-order, undamped, catheter pressure measuring system. Note that there is significant amplification of the signal as the input frequency approaches the natural frequency of the system.

8.3.7   Case 2: the undriven, damped catheter measurement system

A damped system would be one in which the viscosity of the fluid in the catheter, and therefore the corresponding hydraulic resistance in the catheter, are relatively large; therefore significant. A practical example of a damped measuring system, is one in which the viscosity is relatively large compared to the mass of the system. A relatively small diameter catheter, with air bubbles or perhaps with a crimp in the tubing, and with a relatively high viscosity fluid like blood, might behave in this way. A system like this could have a relatively low response time but might

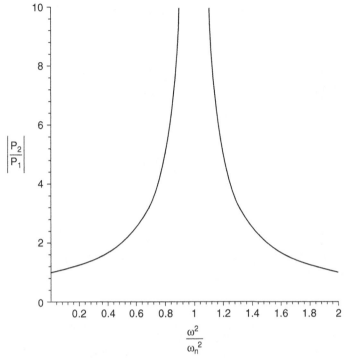

Figure 8.8   Response of a second-order, undamped, pressure measuring system.

provide a kind of filtering action to eliminate higher-frequency noise on the pressure waveform. If the system is undriven, this simply means that pressure may change, as with a step input, but that there is no oscillating pressure input. Imagine the pop test described in Sec. 8.3.7 and shown in Fig. 8.9.

For the undriven system, we set the driving pressure equal to zero and Eq. (8.27) becomes Eq. (8.42).

$$\ell L \ddot{P}_2 + \ell R_v \dot{P}_2 + E P_2 = 0 \tag{8.42}$$

Since the system is undriven, we might imagine that when there is a step change in pressure, for example, going from some positive pressure to zero pressure, that the system would follow an exponential decay. We can try $P_2 = Ae^{\lambda t}$, where λ is 1 divided by the time constant of the system. The first and second derivatives of P_2 now become $\dot{P}_2 = \lambda Ae^{\lambda t}$ and $\ddot{P}_2 = \lambda^2 Ae^{\lambda t}$. By substituting those values into Eq. (8.42) we arrive at Eq. (8.43).

$$\ell L \lambda^2 Ae^{\lambda t} + R_v \ell \lambda Ae^{\lambda t} + E Ae^{\lambda t} = 0 \tag{8.43}$$

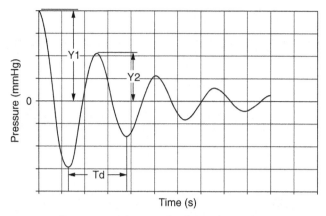

Figure 8.9 Step response for an under damped second-order system.

By using the substitutions suggested in (8.27), $\ell L = m$ and $R_v\ell = b$, we see a simplified version of Eq. (8.43).

$$m\lambda^2 + b\lambda + E = 0 \tag{8.44}$$

It is now possible to use the quadratic formula to solve for lambda in terms of b, M, and volume modulus of elasticity of the transducer E.

$$\lambda = \frac{-b \pm \sqrt{b^2 - 4mE}}{2m} \tag{8.45}$$

Recall from the definition of the damping ratio and critical damping, from Eq. (8.31), that the system is critically damped if it has a damping ratio of 1, which means that $b = b_{\text{critical}} = 2\sqrt{Em}$. By substituting that term into Eq. (8.45), it is possible to solve for the lambda associated with critical damping.

$$\lambda_{\text{critical}} = \frac{-b}{2m} \tag{8.46}$$

Recall that for the catheter measuring system, the general variables in Eq. (8.27) k, m, and b have the following definitions: $k = E$, $m = \ell L$, and $b = R_v\ell$.

$$\lambda_{\text{critical}} = \frac{\ell R_v}{2\sqrt{E\ell L}} \tag{8.47}$$

Note that because we set the damping ratio to one, i.e., when $b = 2\sqrt{Em}$, then Eq. (8.47) yields the same expression for the damping ratio in Eq. (8.33). Again, λ is one over the time constant of the system and ζ is the dimensionless damping ratio.

$$\text{damping ratio} \equiv \zeta = \frac{\ell R_v}{2\sqrt{E\ell L}} \tag{8.33}$$

The output of our system is therefore defined by the expression shown in Eq. (8.48).

$$P_2 = Ae^{\lambda t} + Bte^{\lambda t} = Ae^{-\left(\frac{b}{2m}t\right)} + Bte^{-\left(\frac{b}{2m}t\right)} \tag{8.48}$$

where $b = R_v\ell$ and $m = \ell L$. Substituting $b = 2\sqrt{Em}$ into Eq. (8.48) also yields

$$P_2 = Ae^{-\sqrt{\frac{E}{m}}t} + Bte^{-\sqrt{\frac{E}{m}}t} \tag{8.49}$$

Comparing the definition for the natural frequency from Eq. (8.32), we find that we now have an expression for critically damped catheters that relates the output of the measuring system to the natural frequency of the catheter.

$$P_2 = Ae^{-\omega_n t} + Bte^{-\omega_n t} \tag{8.50}$$

Critically damped systems are a very important case. This kind of system is on the borderline, but does not oscillate as it returns to equilibrium after perturbation. This disturbance is damped out as quickly as possible, as in the case of a door closer designed to close a screen door as quickly as possible with no back and forth motion.

The critical time constant for the system is $1/\lambda_{\text{critical}}$ as shown in Eq. (8.51).

$$\text{critical time constant} = \frac{2m}{b} = \frac{2\ell L}{R_v\ell} = \frac{2L}{R_v} = \frac{c_u r_o^2}{c_v 4v} \tag{8.51}$$

It is useful to point out that Eqs. (8.52) and (8.53) are general expressions for the catheter measuring system and not a special case for the critically damped system.

$$\zeta = \frac{b}{2\sqrt{EM}} = \frac{R_v\ell}{2\sqrt{E\ell L}} \tag{8.52}$$

$$\omega_n = \sqrt{\frac{k}{m}} = \sqrt{\frac{E}{\ell L}} \tag{8.53}$$

8.3.8 Pop test—measurement of transient step response

One way to determine some characteristics of a second-order system is to create a step input to the system and measure the response. One straightforward and simple method of measuring the system response is to input a step change into the system by popping a balloon or surgical glove and measuring the response. Imagine a catheter with one end connected to a pressure transducer and the other end inserted into a container covered with a balloon or surgical glove. If the system is pressurized and the balloon or glove is popped, the resulting pressure output from the system looks like that shown in Fig. 8.9. The original steady state pressure was Y_1 and the pressure output of the transducer oscillates back and forth around zero, depending on the characteristics of the pressure measuring system.

The period of the oscillation is labeled T_D in Fig. 8.9. The damped natural frequency of the system, measured in radians per second, is given by the equation $\omega_D = \frac{2\pi}{T_D}$. The logarithmic decrement is given as δ and is defined by the equation

$$\delta = \ln\left[\frac{P_2(0)}{P_2(1)}\right]$$

The damping ratio of the pressure measuring system can now be calculated from the logarithmic decrement as shown in Eq. (8.54).

$$\zeta = \sqrt{\left[\frac{\delta^2}{4(\pi)^2 + \delta^2}\right]} \qquad (8.54)$$

The relationship between the natural frequency, the damped natural frequency, and the damping ratio is shown in Eq. (8.55).

$$\frac{\omega_D}{\omega_n} = \sqrt{1 - \zeta^2} \qquad (8.55)$$

By measuring the amplitude ratio of successive positive peaks, it is possible to determine the logarithmic decrement of the system. From the logarithmic decrement it then becomes straightforward to measure the damped natural frequency, the damping ratio, and the undamped natural frequency.

If the natural frequency of the catheter measuring system is very low, some relatively low frequency harmonics of the pressure waveform can be amplified, distorting the pressure waveform significantly. If possible, choose a catheter-transducer measuring system with a natural frequency that is 10 times greater than the highest harmonic frequency of interest in the pressure waveform.

8.4   Flow Measurement

Measurement of blood flow is an important indicator of the function of the heart and the cardiovascular system in general. Cardiac output is the measurement of the flow output of the heart and is a measure of the ability of the heart and lungs to provide oxygenated blood to tissues throughout the body. In this chapter we look at several techniques for measuring cardiac output as well as techniques for measuring blood flow in specific vessels as a function of time.

Adolph Fick was born in 1829 in Kassel, Germany. Fick studied medicine, later became interested in physiology, and took a position in Zurich where he was already making contributions to the scientific literature in physics at the age of 26. Fick's first contribution as a physicist was a statement that diffusion is proportional to concentration gradient. Fick is most famous because of a brief, obscure publication in 1870 in which he described how mass balance might be used to measure cardiac output—the Fick principle.

8.4.1   Indicator dilution method

The general formulation of the indicator dilution method of measuring blood flow is one that is based on the Fick principle. The substance is injected into the bloodstream in a known quantity. The flow rate of blood through a given artery, which is also the time rate of change of volume of blood moving through the same artery, can be predicted if the rate of injection of some mass of the indicator is known, along with the change in concentration of the indicator in the blood. For example, imagine that a bolus of indicator dye is injected into the bloodstream. The concentration of the dye at a single point in some vessel could be measured and will be high immediately upon injection and will reduce with time as the dye is first diluted and then cleared from the system. The concentration will be a function of time. For this case change in concentration, ΔC, would be the change in concentration over time at a specific point. (Note that the symbol C has been used to represent compliance earlier in the chapter, and should not be confused with C for concentration).

The general formulation for the flow rate is given by Eq. (8.56), where m is the given quantity of indicator substance injected, dm/dt is the time rate of injection of m, V is the blood volume, and C is the concentration of m in blood. Depending on the specifics of the method used, ΔC may be a change in concentration at a single point in a blood vessel at two different times, or it may be the instantaneous difference in concentration between two different locations. Examples demonstrating the indicator-dilution method are described in more detail in the following sections.

$$\text{Flow} \equiv Q = \frac{dV}{dt} = \frac{dm/dt}{\Delta C} \qquad (8.56)$$

8.4.2   Fick technique for measuring cardiac output

The Fick principle relates the cardiac output Q to oxygen consumption and to the arterial and venous concentration of oxygen as shown in Eq. (8.57).

$$Q = \frac{dV}{dt} = \frac{dm/dt}{\Delta C} = \frac{dm/dt}{C_a - C_v} \qquad (8.57)$$

For Eq. (8.57), Q is the cardiac output in liters per min, dm/dt is the consumption of O_2 in liters per min, C_a is the arterial concentration of O_2 in liters per liter, and C_v is the venous concentration of O_2 in liters per liter. Using a device called a spirometer, it is possible to measure a patient's oxygen consumption. It is also possible to measure both arterial and venous oxygen concentration from a blood sample collected through a catheter.

Oxygen is the indicator used in this method which enters the bloodstream through the pulmonary capillaries. Cardiac output is calculated based on the oxygen consumption rate and the oxygen concentration in arterial and venous blood.

8.4.3   Fick technique example

A patient's spirometer oxygen consumption is 250 mL/min while her arterial oxygen concentration is 0.2 mL/mL and her venous oxygen concentration 0.15 mL/mL. Cardiac output is calculated by dividing the oxygen consumption rate by the change in concentration (0.22−0.15) L/L. The patient's cardiac output is 5 L/min.

$$Q = \frac{dV}{dt} = \frac{dm/dt}{\Delta C}$$

$$Q = \frac{dV}{dt} = \frac{dm/dt}{\Delta C} = \frac{0.25 \text{ liter/min}}{0.20 \text{ liter/liter} - 0.15 \text{ liter/liter}} = 5 \frac{\text{liters}}{\text{min}}$$

8.4.4   Rapid injection indicator-dilution method—dye dilution technique

Indocyanine green is used clinically as an indicator to measure blood flow. This method could be used to measure cardiac output but is also applicable to local flow measurements like cerebral blood flow or femoral arterial flow. A curve is generated using a constant flow pump and blood is continuously drawn from the vessel of interest into a colorimeter cuvette to measure color (dye concentration).

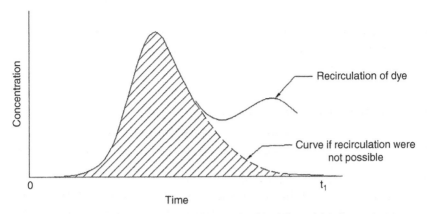

Figure 8.10   Recirculation curve used for measuring blood flow with indocyanine-green dye dilution method. (Reprinted from Webster, JG, ed, *Medical Instrumentation: Application and Design*, 1998, Fig. 8.2, p. 335. With permission from John Wiley & Sons.)

Figure 8.10 shows a dye concentration curve that is generated by this method. A bolus of dye is rapidly injected into the vessel of interest. The solid line in the figure represents the fluctuation in concentration of the dye, in the blood, after injection. The variable m represents the amount of dye injected. Q represents the flow rate and C represents the dye concentration. The area under the curve in Fig. 8.10 represents the amount of dye injected. The flow rate can be calculated as shown in Eq. (8.58).

$$Q = \frac{m}{\int_0^{t_1} C(t)dt} \tag{8.58}$$

In Eq. (8.58) Q is approximately constant and C is a function of time.

8.4.5   Thermodilution

The technique of thermodilution uses heat as an indicator for measuring blood flow thereby avoiding the use of dyes like indocyanine green. Cooled saline can be injected into the right atrium while a thermistor placed in the pulmonary artery measures temperature. Temperature is used as a measure of the concentration of the indicator substance; in this case heat. Flow can then be calculated using Eq. (8.59).

$$Q = \frac{q}{\rho_b c_b \int_0^{t_1} \Delta T_b(t)dt} \quad \frac{\mathrm{m}^3}{\mathrm{s}} \tag{8.59}$$

where Q is the flow rate in m^3/s; q represents heat content of injectate, J; ρ_b represents the density of blood, kg/m^3; c_b represents specific heat of blood, J/(kg°K) = Nm/(kg°K); and finally, ΔT is the temperature change measured in degrees Kelvin, °K.

8.4.6 Electromagnetic flowmeters

A conductor moving through a magnetic field generates an electromotive force in that conductor. The electromotive force (EMF) results in the flow of current which is proportional to the speed of the conductor. Based on this principle, an electromagnetic flowmeter can measure flow of a conducting fluid when that conducting fluid flows through a steady magnetic field. Figure 8.11 shows a schematic of a blood vessel between permanent magnets that generate a magnetic field within the blood vessel. A current is generated, between the two electrodes shown, that is proportional to the average blood velocity across the cross section of the vessel. The EMF generated between the electrodes placed on either side of the vessel is given in Eq. (8.60).

$$e = \int_0^{L_1} \vec{u} \times \vec{B} \cdot d\vec{L} \tag{8.60}$$

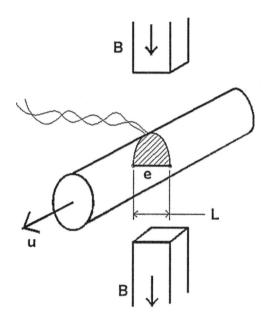

Figure 8.11 Schematic of an electromagnetic flow meter with magnetic field $\vec{B}$ and blood velocity $\vec{u}$. (Adapted from Webster, JG, ed., *Medical Instrumentation: Application and Design*, John Wiley & Sons, New York, 1998, Fig. 8.3.)

where $\vec{u}$ represents the instantaneous blood velocity, m/s
$\vec{L}$ represents the length between electrodes, m
$\vec{B}$ represents the magnetic flux density, T
and T has the units Weber/m^2 $= \dfrac{\text{Volt-second}}{\text{meter}^2}$

The variable e represents the EMF or voltage generated in response to blood velocity.

For a uniform magnetic flux density B and a uniform velocity profile, the expression may be simplified to the scalar Eq. (8.61).

$$e = BLu \qquad (8.61)$$

Some potential sources of error in electromagnetic flowmeters include

1. Current flows from high to low velocity areas.

2. Current can be shunted through the vessel wall. This effect varies with hematocrit.

3. Fluid outside the vessel wall can shunt current.

4. Nonuniform magnetic flux density will cause a variation in the flowmeter output, even at a constant flow rate.

One typical type of electromagnetic probe is the toroidal type cuff-perivascular probe. The toroidal cuff uses two oppositely wound coils and a Permalloy core. The probe should fit snug around the vessel and this requires some constriction during systole.

8.4.7 Continuous wave ultrasonic flowmeters

Another type of flowmeter used in medical applications is the continuous wave Doppler flowmeter. This type of flowmeter is based on the Doppler principle when a target recedes from a fixed sound transmitter, the frequency is lowered because of the Doppler effect. For relatively small frequency changes, the relationship in Eq. (8.62) is true.

$$\frac{F_d}{F_o} = \frac{u}{c} \qquad (8.62)$$

In Eq. (8.62), F_d is the Doppler shift frequency, F_o is the source frequency, u represents the target velocity in this case an erythrocyte, and c represents the velocity of sound. For two shifts, one from source and one to target:

$$\frac{F_d}{F_o} \approx \frac{2u}{c} \qquad (8.63)$$

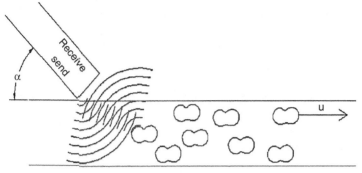

Figure 8.12 Doppler continuous wave probe elevated at an angle α with respect to the patient's skin.

Consider now a probe at an angle α with respect to the patient's skin, as shown in Fig. 8.12. The signal can now be calculated as:

$$F_d = \frac{2f_o u \cos \alpha}{c} \tag{8.64}$$

Some problems that occur with continuous wave Doppler ultrasound are as follows:

1. The Doppler shifted frequency is actually not a single frequency, but a mixture of many frequencies.
2. The velocity profile of the flow field is not constant across the vessel cross section.
3. The beam of the sound wave is broad and spreading.
4. Turbulence and tumbling of cells also cause a Doppler shift.
5. Simple meters do not measure reverse flow.

8.4.8 Example problem—continuous wave Doppler ultrasound

Using a continuous wave Doppler with a carrier frequency of 7 MHz, $\alpha = 45°$, blood velocity of 150 cm/s and speed of sound = 1500 m/s, find the Doppler shifted frequency. Is it in the audible range?

$$F_d = \frac{2f_o u \cos \alpha}{c}$$

$$F_d = \frac{2(7 \times 10^6 \,\text{Hz})\left(1.5\,\frac{m}{s}\right)\cos(45)}{\left(1500\,\frac{m}{s}\right)} \approx 10 \text{ kHz}$$

Yes. The frequency 10 kHz is in the audible range of a normal, healthy human.

8.5   Summary and Clinical Applications

Chapter 8 introduces flow and pressure measurement schemes. A very quick review to remind the reader of the important clinical applications of those measurements is given below.

Indirect pressure measurements   According to the American Heart Association, high blood pressure is listed as a primary or contributing cause of death in more than a quarter of a million Americans each year. Nearly one in three U.S. adults has high blood pressure and 30 percent of those people do not know it. Indirect blood pressure measurement using a sphygmomanometer is the method of choice for measuring blood pressure in the doctor's office and is critical to the management regimen of this disease.

Intravascular pressure measurement   Cardiologists sometimes insert a strain-gauge-tipped pressure catheter into an artery to determine whether a specific stenosis is the cause of decreased blood flow. If the pressure is much lower downstream from the blockage than the pressure immediately upstream, this indicates that the lesion is the cause. This procedure can also be used to evaluate the effectiveness of catheterization and stenting. This use of intravascular pressure measurement is an important aspect of the diagnosis of coronary arterial disease.

Extravascular pressure measurement   Pulmonary capillary wedge pressure is the pressure measured by inserting a balloon-tipped catheter from a peripheral vein into the right atrium, through the right ventricle, and then positioned within a branch of the pulmonary artery. Measurement of pulmonary capillary wedge pressure (PCWP) gives an indirect measure of left atrial pressure and is particularly useful in the diagnosis of left ventricular failure and mitral valve disease.

The catheter used for this procedure has one opening (port) at the tip of the catheter and a second port several centimeters proximal to the balloon. These ports are connected to extravascular pressure transducers allowing the pressure measurement.

Cardiac output measurement   Cardiac output is one of the main determinants of organ perfusion. During a coronary artery bypass, dislocation of the heart changes the cardiac output and so it is important for the anesthesiologist to have a reliable tool for assessing hemodynamic status to avoid situations with disastrously low cardiac output. There are a number of methods used to assess cardiac output, including the Fick technique, rapid injection dye dilution, and thermodilution.

Electromagnetic and continuous wave Doppler flowmeters   These two devices are two tools in the physicians' arsenal that can be used to measure blood flow in a variety of arteries. For example, physicians use Doppler flowmeters to diagnose peripheral arterial disease. The physician measures flow waveforms and could compare, for example, the waveforms from arteries in the left arm to those from the right arm. A significant difference in the waveforms between the two vessels could indicate a stenosis in one of the vessels.

While electromagnetic flowmeters are more invasive, they are commonly used as the gold standard for measuring blood flow in animal experiments in laboratory settings. In Chap. 10, a model of flow through the mitral valve in the porcine model is presented. That model was developed from data collected at the University of Heidelberg and the flow measurements were made using the electromagnetic flowmeter.

Review Problems

1. A pressure transducer is used with a 0.92 mm I.D. × 100 cm long catheter, and the transducer has a volume modulus of elasticity, $E = 0.40 \times 10^{15}$ N/m^5. The system has a damping ratio of approximately 30% of critical when filled with an unknown fluid (use $\rho = 1000$ kg/m^3). Used values of $C_U = 1.1$ and $C_v = 1.6$.

   (a) Find the natural frequency of the transducer/catheter system.
   (b) Find the damped natural frequency of the system.

2. A liquid-filled catheter, 100 cm long, with an internal radius of 0.033 cm, is connected to a strain gauge pressure transducer. From a pop test, the logarithmic decrement was found to be 1.2, and the damped frequency was 25 Hz. Find the undamped natural frequency. Do you think for this combination of pressure, transducer and catheter would be a suitable system for measuring pressure variations in which the highest frequencies of interest are 10 Hz?

3. From a pop test of a strain gauge pressure transducer connected to a catheter, the pressure time recording shown below was obtained. Determine the damped natural frequency, the damping ratio, and the natural frequency of the system.

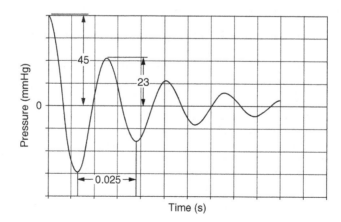

4. A pressure transducer is connected to a relatively rigid catheter. The catheter is filled with liquid. From a pop test:

$P_2(0) = 60$ mmHg
$P_2(1) = 15$ mmHg
$T_d = 0.030$ s

(a) Determine the undamped natural frequency (Hz).
(b) The length of the catheter is 1 m. The inside diameter is 2 mm. Use density of the fluid = 1000 kg/m^3 and $c_u = 1.1$ to estimate the volume modulus of elasticity of the transducer.

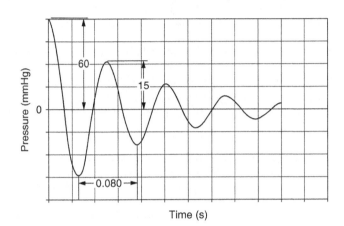

Time (s)

5. A patient's cardiac output is 5500 mL/min while his arterial oxygen concentration is 0.2 mL/mL and his venous oxygen concentration is 0.15 mL/mL. Find this person's spirometer oxygen consumption.

6. Using a continuous wave Doppler with a carrier frequency of 7 MHz, $\alpha = 45°$, the speed of sound = 1500 m/s, and a Doppler shifted frequency of 5000 Hz. find the blood velocity.

7. The peak average velocity (averaged across the cross section) in a dog's aorta, 15 mm in diameter, is 1 m/s. An electromagnetic flowmeter is used with a magnetic flux density of 0.025 tesla. What is the voltage generated at the electrode?

8. Using the thermodilution method for measuring cardiac output, 10 mL of injectate is injected over 2.5 s. The cardiac output is 4.0 L/min. Use the following data to estimate the value of $\int_0^{t_1} \Delta T_b dt$

volume of injectate = 10 mL
dt of injectate = -30 K
density of injectate = 1005 kg/m^3
heat capacity of injectate = 4170 J/(kg K)
density of blood = 1060 kg/m^3
heat capacity of blood = 3640 J/(kg K)

9. Consider the indirect measurement of blood pressure in an obese patient using a sphygmomanometer. Does indirect measurement of blood tend to overestimate or underestimate the actual blood pressure in the obese patient? Is there an inaccuracy in the measurement associated with "hardened arteries" in elderly patients?

10. A rigid-walled catheter is connected to a pressure sensor. The catheter is 1 m long, with an inside diameter of 1.1 mm and filled with water at 20°C. The volume modulus of elasticity of the pressure sensor diaphragm is 0.49×10^{15} N/m^5. Find the natural frequency of the catheter, the damped natural frequency of the catheter, and the damping ratio. Use viscosity of the water = 0.89 cP and $c_v = 1.8$.

11. For the same pressure measuring catheter from Prob. 8.10, add an air bubble to the system. Use an isothermal $\frac{\Delta V}{\Delta P}$ for air of 0.010 $\frac{\text{mL}}{\text{Pa}}$ per liter of air volume. The catheter is 1 m long, with an inside diameter of 1.1 mm and filled with water at 20°C. The volume modulus of elasticity of the pressure sensor diaphragm is 0.49×10^{15} N/m^5. Find the length of air bubble in the catheter required for a damping ratio, ζ, of 0.14.

12. For the same pressure measuring catheter from Prob. 8.10 without the air bubble a 15-cm-long pinch reduces the inside diameter of the catheter to 0.3 mm. Find the damped natural frequency of the catheter with the pinch. Use a c_u of 1.1 and c_v of 1.8.

Bibliography

Cromwell, L, Weibell, FJ, Pfeiffer, EA. *Biomedical Instrumentation and Measurements*, 2nd ed, Prentice-Hall, Englewood Cliffs, NJ, 1980.

Khandpur, RS. *Biomedical Instrumentation. Technology and Applications*, McGraw-Hill, New York, 2005.

Kothari, N, Amaria, T, Hegde, A, Mandke, A, Mandke, NV. "Measurement of Cardiac Output: Comparison of Four Different Methods," *IJTCVS*, 19, 163–168, 2003.

Özyurda, U, Akar, R, Uymaz, O, Oguz, M, Ozkan, M, Yildirim, C, Aslan, A, RTasoz, R. "Early Clinical Experience with the On-X Prosthetic Heart Valve," *Interact. Cardiovasc. Thorac. Surg.*, 4, 588–594, 2005; originally published online Sep 16, 2005.

Pickering, TG, Hall, JE, Appel, LJ, Falkner, BE, Graves, J, Hill, MN, Jones, DW, Kurtz, T, Sheps, SG, Roccella, EJ. "Recommendations for Blood Pressure Measurement in Humans and Experimental Animals," *Hypertension*, 45, 142, 2005.

Szabo, G, Soans, D, Graf, A, Beller, C, Waite, L, Hagl, S. "A New Computer Model of Mitral Valve Hemodynamics during Ventricular Filling," *Eur. J. Cardiothoracic Surg.*, 26, 239–247, 2004.

Togawa, T, Tamura, T, and Öberg, P. *Biomedical Transducers and Instruments*, CRC Press, Boca Raton, FL, 1997.

Webster, JG, ed. *Medical Instrumentation Application and Design*, 3rd ed, John Wiley & Sons, New York, 1998.

Modeling

9.1 Introduction

Models are used widely in all types of engineering, and especially in fluid mechanics. The term **model** has many uses, but in the engineering context, it usually involves a representation of a physical system, a **prototype**, that may be used to predict the behavior of the system in some desired respect. These models can include physical models that appear physically similar to the prototype or mathematical models, which help to predict the behavior of the system, but do not have a similar physical embodiment. In this chapter, we develop procedures for designing models in a way to ensure that the model and prototype will behave similarly.

Consider the question: "Why can a 1.5-mm-long flea fall a meter without injury but a 1.5-m-tall man can't fall 1 km unhurt?" A model is a representation of a physical system that may be used to predict the behavior of the system in some desired respect. The implied question in the stated question above is, "Can the flea be used to model the man?"

It is true that a 1.5-mm-long flea can jump about 1 m. If we use geometric scaling, we might predict that a 1.5-m-long person would jump about 1 km. Since that is clearly not true, then what is wrong with this picture? A careful look at the theory of models will help answer our question.

In Sec. 9.7, we attempt to use a flea to model a man (or vice versa) to model the terminal velocity of the prototype using the model. We will learn, by using the theory of models and dimensional analysis, why this model does not work very well.

9.2   Theory of Models

The **theory of models** is developed from the principles of dimensional analysis. These principles tell us that any prototype can be described by a series of dimensionless parameters, which I will call **Pi terms** in this chapter. For the prototype system, the parameter that we would like to measure, Π_1, can be represented as a function Φ of a set of n dimensionless Pi terms.

$$\Pi_1 = \Phi \, (\Pi_2, \Pi_3, \Pi_4, \ldots, \Pi_n) \tag{9.1}$$

If the relationship Φ between Pi terms describes the behavior of the system, then it would also be possible to develop a similar set of Pi terms for a model that has the same dimensional relationships.

$$\Pi_{1m} = \Phi \, (\Pi_{2m}, \Pi_{3m}, \Pi_{4m}, \ldots, \Pi_{nm}) \tag{9.2}$$

9.2.1   Dimensional analysis and the Buckingham Pi theorem

Edgar Buckingham (1867–1940) was an American physicist who first generated interest in the idea of dimensional analysis. In 1914, Buckingham published the article, "On Physically Similar Systems: Illustration of the Use of Dimensional Equations." The Buckingham Pi theorem states:

> The number of independent dimensionless quantities required to describe a phenomenon involving k variables is n, where $n = k - r$, and where r is the number of basic dimensions required to describe the variable.

To demonstrate the use of the Buckingham Pi theorem, imagine that we would like to perform a test that describes the pressure drop per length of pipe as a function of other variables that affect the pressure gradient (Fig. 9.1).

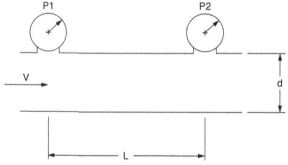

Figure 9.1   Pressure gauges at two points along a pipeline of diameter d, with a mean velocity of the pipe of V.

The first step in the process is to choose three fundamental dimensions, which describe mechanical properties. One might choose from a list of fundamental dimensions like force, length, time, mass, temperature, charge, voltage, and the list would be extensive. In fact, it is pretty clear that charge, voltage, and temperature are not important in this problem, so we will choose force, length, and time. We will designate those dimensions by the characters F, L, and T representing force, length, and time, respectively.

The second and arguably most difficult step in the modeling process will be to determine all important variables. In some models, it will not be obvious which variables are important. Occasionally, to simplify a model, we will begin with a relatively small set of variables and when the model does not suitably represent the prototype we may find out that we need to add variables. For this example problem, we will assume that the pressure gradient $\dfrac{\Delta p}{L}$ is related to average velocity in the pipe, V, the diameter of the pipe, d, the viscosity of the fluid flowing in the pipe, μ, and the density of the fluid, ρ.

$$\Delta p/L = f\,(V, d, \mu, \rho) \tag{9.3}$$

The third step in the process is to write the dimensions of each variable. For example, the variable $\Delta p/L$ has the dimensions of (force/length^3) as shown next.

$$\Delta p/L \sim F/L^3 \tag{9.4}$$

The other variables in the example have dimensions as shown in Eqs. (9.5) through (9.7):

$$V \sim L/T \tag{9.5}$$

$$\mu \sim FT/L^2 \tag{9.6}$$

$$\rho \sim M/L^3 = FT^2/L^4 \tag{9.7}$$

Now that all of the dimensions have been described, we can use Buckingham's Pi theorem to determine the number of required Pi terms to describe this model.

$$n = k - r = 5 \text{ variables} - 3 \text{ basic dimensions} = 2 \text{ Pi terms} \tag{9.8}$$

The result of this analysis is that we have confirmed the following fact. To conduct an experiment to determine the relationship between $\Delta P/L$ and the other important variables, we need to measure only two dimensionless Pi terms and not five different variables.

Pi terms are dimensionless terms, and in this case, it would be possible to choose the following two Pi terms to describe the system.

$$\Pi_1 = (\Delta p/L)d/(\rho V^2)$$

and

$$\Pi_2 = \rho V d/\mu$$

In fact, any two independent Pi terms would suffice. The inverse of either of the above Pi terms would also be valid Pi terms, for example, $1/\Pi_1$ or $1/\Pi_2$. The most important criterion of the two Pi terms is that they are independent terms. Another important factor in choosing the best Pi terms for the problem being studied is that it is important to choose one Pi term that includes the dependent variable to be investigated and that it appears only in that term.

By using **dimensional analysis**, we have now learned that to run this experiment, we do not need to vary density, diameter, or viscosity. It is possible to vary only velocity and measure only pressure gradient. Not only have we reduced the variables from five to two, but we have also created dimensionless terms so that our results are independent of the set of units that we choose. Figure 9.2 shows a series of experiments that compare the pressure gradient, designated ΔP_ℓ in these charts, to the four variables V, d, ρ, and μ. These four comparisons require four separate graphs. This means that we will be able to generate the same amount of data from one experiment that we would have needed four experiments to generate, had we not used dimensional analysis.

Figure 9.3 shows a single dimensionless plot that contains the same information included in all four plots in Fig. 9.2. The plot shows the pressure gradient as a function of V, d, ρ, and μ.

9.2.2 Synthesizing Pi terms

In some models, Pi terms are relatively easy to determine. It is useful here to discuss an algorithm for generating independent Pi terms from a list of variables. To generate a set of independent Pi terms for this example problem, begin by writing down all of the variables, each raised to the power of an unknown as shown below.

$$(\Delta P/L)^a \quad (V)^b \quad (d)^c \quad (\mu)^e \quad (\rho)^f$$

Next, write the dimensions of those variables, in the same format.

$$(F/L^3)^a \quad (L/T)^b \quad (L)^c \quad (FT/L^2)^e \quad (FT^2/L^4)^f$$

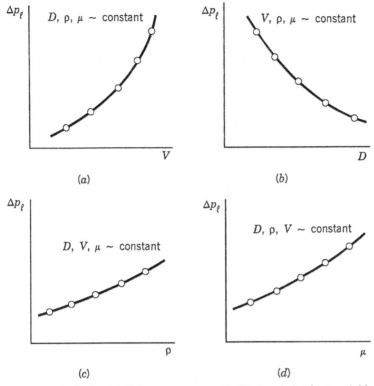

Figure 9.2 Relationship between pressure gradient and the four variables: velocity, pipe diameter, fluid density, and fluid viscosity. (From Munson, BR, et al., *Fundamentals of Fluid Mechanics*, © 1994; John Wiley & Sons, New York. Reprinted with permission of John Wiley & Sons, Inc.)

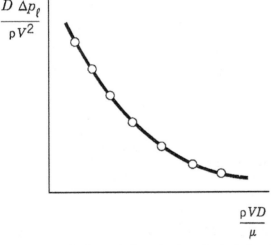

Figure 9.3 Single dimensionless plot that also includes all of the information included in the four plots in Fig. 9.2. (From Munson, BR, et al., *Fundamentals of Fluid Mechanics*, © 1994; John Wiley & Sons, New York. Reprinted with permission of John Wiley & Sons, Inc.)

Now it is possible to write a set of three equations, one for each basic dimension, and to solve them simultaneously for three values. Since we have five unknowns and only three equations, it is necessary to assume the values of two convenient variables.

$$F: \quad a + 0 + 0 + e + f = 0$$

$$L: \quad -3a + b + c - 2e - 4f = 0$$

$$T: \quad 0 - b - 0 + e + 2f = 0$$

For our example, we will assume $a = 1$ and $e = 0$ to produce two independent Pi terms. That will ensure that $\Delta P/L$ will appear in only one Pi term. Solving the three equations yields the following:

$$1 + f = 0 \quad f = -1$$

$$-3 + b + c - 4f = 0 \quad c = 1$$

$$-b + 2f = 0 \quad b = -2$$

Therefore, by using the values of $a = 1$ and $e = 0$, one possible Pi term follows:

$$\Pi_1 = (\Delta P/L)^1 \, V^{-2} d^1 \rho^{-1}$$

To generate the second independent Pi term, let $e = 1$ and $a = 0$. This will ensure that every variable appears in one of the Pi terms and that $\Delta P/L$ does not appear in Π_2.

$$1 + f = 0 \quad f = -1$$

$$b + c - 2 - 4f = 0 \quad c = -1$$

$$-b + 1 + 2f = 0 \quad b = -1$$

$$\Pi_2 = V^{-1} d^{-1} \mu \rho^{-1} = \mu / \rho V d$$

Note that Π_2 is the reciprocal of the Reynolds number, a well-known dimensionless parameter used in many fluid mechanics applications.

Therefore, if we make Π_2 for the model equal to Π_2 for the prototype, then Π_1 for the model will predict meaningful values for Π_1 of the prototype.

9.3 Geometric Similarity

When we equate Pi terms involving length ratios only, the model satisfies the condition of **geometric similarity**. Scale models in which all three dimensions of the model (height, width, and length) have the same scale have geometric similarity. A common dimensionless Pi term involving geometry is the ratio of length to width or the ratio of two lengths.

$$L_1/L_2 = L_{1m}/L_{2m}$$

If a model is a 1/10 scale model, this means that the ratio of lengths between the model and the prototype is 1/10. The flea in our example introduced above might be considered a 1/1000 **scale model** of a human, if it were similarly proportioned to the human. If we only know the flea's length and the man's height, we can consider the flea geometrically similar. We will call the flea a 1/1000 scale model for a first approximation.

But if we consider the overall effectiveness of the model, what about the other Pi terms that have forces and time in them? For a complete model, we also need some other types of similarity.

9.4   Dynamic Similarity

For a perfect model, we must be sure that we have included all important variables. Further, all Pi terms in the model must be equal to each Pi terms in the prototype. When we achieve geometric similarity by matching the geometric Pi terms, as with a scale model airplane, the model will look like a smaller version of the prototype but will not necessarily behave like the prototype. When we equate Pi terms involving force ratios we achieve dynamic similarity.

For example, density is a variable that contains force as one of its dimensions. A Pi term in some fluid mechanics problems that includes drag force and density could be the term that follows. For this Pi term the variable D represent drag force, ρ represents fluid density, V represents fluid velocity, and t_1 and t_2 are geometric dimensions.

$$\Pi_1 = [D/(\rho V^2 t_1 t_2)]$$

To achieve **dynamics similarity**, Π_1 in the prototype must equal Π_1 in the model.

$$[D/(\rho V_{max}^2 t_1 t_2)]_p = [D/(\rho V_{max}^2 t_1 t_2)]_m$$

9.5   Kinematic Similarity

When we equate Pi terms involving velocity or acceleration ratios we obtain kinematic similarity. Many Pi terms that have a force dimension will also have a velocity or acceleration dimension. For example, the Reynolds number is a common Pi term in fluid mechanics that has both

TABLE 9.1   Shows Some Common Dimensionless Parameters

Pi term	Name	Ratio
$\rho VL/\mu$	Reynolds number, Re	Inertia/viscous
$V/(gL)^{1/2}$	Froude number, Fr	Inertia/gravity
$p/(\rho V^2)$	Euler number, Eu	Pressure/inertia
V/c	Mach number, Ma	Velocity ratio
$L(\omega\rho/\mu)^{1/2}$	Womersley number, α	Transient inertial forces/viscous forces
$\rho V^2 L/\sigma$	Weber number, We	Inertia/surface tension

force dimensions and velocity dimensions. The force dimension shows up in both density and viscosity. To achieve kinematic similarity, and to achieve dynamics similarity, the model and the prototype must have Reynolds number similarity. That is, the Reynolds number of the model must equal the Reynolds number of the prototype.

$$(\rho VD/\mu)_p = (\rho VD/\mu)_m$$

To achieve similitude, we must have a model with geometric similarity, **dynamic similarity**, and **kinematic similarity** when compared to our prototype.

9.6   Common Dimensionless Parameters in Fluid Mechanics

Dimensionless Pi terms are used in many fluid mechanics applications. Some of the terms are used relatively frequently and have names. Some important dimensionless groups in fluid mechanics are shown in Table 9.1. In the table ρ is density, V is fluid velocity, and L is some geometric length. Viscosity is represented by μ and gravitational acceleration is represented by g. Pressure is denoted by p. The frequency of the changing flow is ω. Surface tension in the Weber number is represented by σ.

9.7   Modeling Example 1—Does the Flea Model the Man?

If we use a flea to model a man (or vice versa) we need to decide on a few important parameters. Let us first assume that we want to model the terminal velocity of the prototype using the model. The terminal velocity is the maximum velocity that a falling object in free fall will reach. We could assume that the maximum velocity of the prototype will be a function of drag D; the weight of the object W; the geometry of the object (which we will designate by three dimensions L, t_1, and t_2); the density of the fluid in which the object is falling ρ; and the viscosity of the fluid in which the object is falling μ. Mathematically, we can write the relationship as follows:

$$V_{max} = f(D, W, L, t_1, t_2, \rho, \mu)$$

Therefore, for this modeling problem, there are eight variables and three basic dimensions that are required to describe the units on those variables. The three basic dimensions are time, length, and force. If our assumptions are valid, then the required number of Pi terms is $n = 5$.

$$n = k - r = 8 - 3$$

I will choose the following five independent Pi terms:

$\Pi_1 = \rho V_{max} L / \mu$ Reynolds number

$\Pi_2 = D / (\rho V_{max}^2 t_1 t_2)$ Coefficient of drag

$\Pi_3 = D / W$ Drag Force to Weight ratio

$\Pi_4 = t_1 / L$ Geometrical scale

$\Pi_5 = t_2 / L$ Geometrical scale

For a first approximation for our model, we will take the flea weight to be 10^{-4} g. The flea length will be 1.5 mm and we will consider the flea and the man to have the same aspect ratio. That means that we will assume Π_4 and Π_5 to be identical in the model and prototype a priori. The man's length will be 1.5 m and the weight will be 100 kg.

Starting from this data, our geometric scale is 1000:1. The man is 1000 times longer and 1000 times wider than the flea. Because we are looking to model the terminal velocity of the falling man and/or the flea, the drag force will be equal to weight for $\Pi_3 = 1$.

For kinematic and dynamic similarity let $\Pi_{1_{man}} = \Pi_{1_{flea}}$ and let $\Pi_{2_{man}} = \Pi_{2_{flea}}$. The resulting equations are shown below.

$$\Pi_2 = [W / (\rho V_{max}^2 t_1 t_2)]_{man} = [W / (\rho V_{max}^2 t_1 t_2)]_{flea}$$

$$(V_{man} / V_{flea})_2 = (W_{man} / W_{flea})(t_1 * t_2)_{flea} / (t_1 * t_2)_{man}$$

$$(V_{man} / V_{flea})_2 = (10^5 / 10^{-4})(1/1000)^2 = 1000$$

If the second Pi term is identical in the flea and the man, the terminal velocity of the man will be ~30 times that of the flea.

$$\Pi_1 = (\rho V_{max} L / \mu)_{man} = (\rho V_{max} L / \mu)_{flea}$$

$$(V_{man} / V_{flea}) = L_{flea} / L_{man} = 1/1000$$

If the first Pi term, the Reynolds number, is identical in the flea and the man, then the velocity of the man will be 1000 times that of the flea. We can see that these two conditions are mutually exclusive and the man and the flea, falling through the same fluid, are not a good model for each other. For Reynolds similarity, the man must fall 1000 times slower. For drag similarity, the man needs to fall 30 times faster.

If you want to allow the man and flea to fall at the same velocity and have Reynolds number similarity, it would also be theoretically possible to change fluids so that we have a different kinematic viscosity, v. We might ask ourselves whether it is practical. The kinematic viscosity of standard air is $v_{air} = 1.46 \times 10^{-5}$ m^2/s. Now if we set the velocity of the man and the flea equal, with the constraint of maintaining Reynolds number similarity, the equation is shown below.

$$\frac{\left(\frac{\rho}{\mu}\right)_{man}}{\left(\frac{\rho}{\mu}\right)_{flea}} = \frac{l_{flea}}{l_{man}} = \frac{1}{1000}$$

Since viscosity divided by density is equivalent to kinematic viscosity; $\mu/\rho = v$, and if the flea is falling through air and the man is falling through an alternate fluid, it is also possible to write the equation as:

$$\frac{v_{air}}{v_{alternate}} = \frac{1}{1000}$$

For Reynolds number similarity the man needs to fall through a fluid that has a kinematic viscosity 1000 times greater than that of standard air or ~1.5×10^{-2} m^2/s. Glycerine at 20°C has a kinematic viscosity of 1.2×10^{-3} m^2/s, so we would need to use glycerine at a much cooler temperature to achieve that kinematic viscosity! Because of glycerine's very large density compared to air, now buoyancy would become a significant factor and we realize that the flea just is not a good model for the man (and vice versa).

Our original question was, "Why can a 1.5 mm flea can fall a meter without injury but a 1.5 m man can't fall 1 km unhurt?" The answer is that the model and prototype are geometrically similar (1/1000 scale) but not dynamically and kinematically similar. The low Reynolds number and relatively high drag effect predict that the flea falls very slowly!

9.8 Modeling Example 2

We would like to study the flow through a 5-mm-diameter venous valve carrying blood at a flow rate of 120 mL/min. We will use water instead of blood, which is more difficult to obtain and more difficult to work with. Take the viscosity of blood to be 0.004 Ns/m^2 and the viscosity of water to be 0.001 Ns/m^2. Complete geometric similarity exists between the model and prototype. Assume a model inlet of 5 cm in diameter. Determine the required flow velocity in the model that would be required for Reynolds number similarity.

Begin by calculating the given mean velocity in the prototype. The mean velocity in the prototype can be calculated from the flow rate in the prototype divided by the cross-sectional area in the prototype.

$$\frac{Q_p}{A_p} = \frac{Q_p}{\pi r_p^2} = \frac{2 \dfrac{cm^3}{s} \dfrac{m^3}{10^6 cm^3}}{\pi(2.5/1000)^2 m^2} = 0.1018 \frac{m}{s} = 10.2 \frac{cm}{s}$$

Next, recognize that for Reynolds number similarity, the following equation must be true, where the subscript p represent the Reynolds number in the prototype and the subscript m represents the Reynolds number in the model.

$$\left(\frac{\rho VD}{\mu}\right)_p = \left(\frac{\rho VD}{\mu}\right)_m$$

$$\bar{V}_m = \frac{\mu_m}{\mu_p} \frac{\rho_p}{\rho_m} \frac{D_p}{D_m} \bar{V}_p = \frac{1}{4} \frac{1060}{1000} \frac{5}{50} \bar{V}_p = 0.0265 \, \bar{V}_p$$

$$\bar{V}_m = 0.0265 \, \bar{V}_p = 10.18 \text{ cm/s} \times 0.0265 = 0.270 \frac{cm}{s}$$

The velocity in the model should be 0.27 cm/s to achieve Reynolds number similarity to the prototype. The much slower velocity required in the model is a result of the larger diameter in the model when compared to the diameter of the prototype.

9.9 Modeling Example 3

The velocity V of an erythrocyte settling slowly in plasma can be expressed as a function of diameter d, thickness t, plasma viscosity μ_p, erythrocyte density ρ_e, plasma density ρ_p, and gravity g.

Written in mathematical formulation, this equals to

$$V_s = f(d, t, \mu \, \rho_e, \rho_p, g)$$

Determine a suitable set of Π terms for expressing this relationship. First, write the dimensions of each variable in the equation above.

$$V_s = L/T \qquad d = L \qquad t = L$$

$$\mu = FT/L^2 \qquad \rho_e = \rho_p = M/L^3 \qquad g = L/T^2$$

Next, find the number of Pi terms:

$$n = k - r = 7 - 3 = 4 \text{ required Pi terms}$$

The Pi terms could be:

$$\pi_1 = t/d \quad \pi_3 = \rho_p V_s d/\mu \quad \text{(Reynolds number)}$$

$$\pi_2 = \rho_e/\rho_p \quad \pi_4 = V_s^2/gd \quad \text{(Froude number)}^2$$

Once the important relationships have been determined, it would be possible, for example, to plot the Froude number as a function of Reynolds number for a given thickness/diameter ratio and for a given erythrocyte versus plasma ratio. As long as the Pi terms in the model are equal to the Pi terms in the prototype, that plot would be equally valid for both prototype and model. Figure 9.4 shows a plot of the Reynolds number versus the

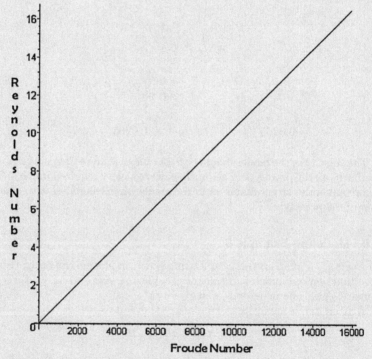

Figure 9.4   Relationship of Reynolds number versus Froude number for the erythrocyte settling problem.

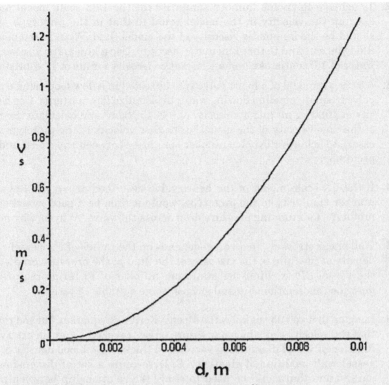

Figure 9.5   Plot of particle sedimentation versus diameter.

Froude number for a density of plasma of 1000 kg/m^3 and a density of erythrocytes of 1200 kg/m^3, a viscosity of 0.0035 Ns/m^2, g = 9.81 m/s^2, and for a thickness-to-diameter ratio of 25. Note that this plot has no units, and as long as the Pi terms are equal in the model and the prototype, the plot is valid.

Figure 9.5 shows the erythrocyte settling rate as a function of cell diameter for a density of plasma of 1000 kg/m^3 and a density of erythrocytes of 1200 kg/m^3, a viscosity of 0.0035 Ns/m^2, g = 9.81 m/s^2, and for a thickness-to-diameter ratio of 25. Note that this plot has units associated and is thus valid only for this set of parameters.

Review Problems

1. Consider a 1/32 scale model airplane which "looks" very similar to the prototype airplane being modeled. What type of similarity does this model have? Would you expect it to fly? If not, describe the other types of similarity you would need for the airplane model to behave similarly to the prototype.

2. To achieve Reynolds number similarity for the 1/32 scale model above, keeping the velocity in the model equal to that in the prototype, what should be the kinematic viscosity of the model fluid. What is the nearest fluid you can find to that kinematic viscosity. Use a kinematic viscosity for water of 1.0 centistoke and a kinematic viscosity for air of 17 centistokes.

3. A 5× scale model of a heart valve is to be tested in a flow loop using a clear polycarbonate pipeline flowing water to visualize flow patterns. Use a density of 1060 kg/m^3 and a viscosity of 0.0045 Ns/m^2 and calculate the ratio of the flow velocity of the model to the flow velocity of the prototype necessary to achieve Reynolds number similarity between the model and the prototype.

4. If the 5× scale model of the heart valve described above has Reynolds number similarity to the prototype, would it then be a good model of the prototype for studying pressure drop across the valve? Why or why not?

5. Wall shear stress τ_w, in a vessel depends on the radius of the vessel r, the density of the fluid ρ, the viscosity of the fluid μ, the pressure gradient in the vessel dP/dx. Find the minimum number of Pi terms required to describe this relationship and suggest some suitable Pi terms.

6. Imagine that you do not know the Moens–Korteweg equation, but you realize that the velocity c at which a pressure pulse travels through an artery is a function of vessel diameter d, vessel wall thickness t, blood density ρ, and vessel wall modulus of elasticity E. Determine a set of dimensionless parameters that could be used to study the relationship between pulse wave velocity and the wall thickness of the vessel.

7. When Poiseuille first developed Poiseuille's law relating flow in a pipe Q to pressure drop along a pipe $\Delta P/l$ he developed the law empirically. Using dimensional analysis, generate a set of Pi terms and suggest an experiment to determine Poiseuille's law for flow through a rectangular channel.

8. You are designing an experimental system to model pressure drops in Poiseuille flow in a 1-mm-diameter artery. You would like to use water in your model through a tube that has a diameter of 1 cm. What velocity and flow rate should be used in the model? Pressure drop in the artery is represented by $\Delta P/L$, blood density by ρ, blood velocity by V, vessel diameter by D, and blood viscosity by μ. The velocity you would like to test in the prototype artery is 5 cm/s. Use a blood viscosity of 3.5 centipoise and blood density of 1060 kg/m^3. Use a viscosity of water of 0.7 centipoise and density of 1000 kg/m^3.

9. For the solution in Prob. 9.8, write the names of the Pi terms using the names of dimensionless parameters from Table 9.1.

10. Two pipelines have equal diameters. Water flows in one pipeline and air flows in the second. Temperature is constant and equal in both pipelines.

You can control the average flow velocity which remains equal in both pipelines. Use a ratio of dimensionless parameters to determine which flow will become turbulent first, the water or the air, as you increase velocity in both pipes?

11. Two pipelines have equal diameters. Water flows in one pipeline and air flows in the second. Temperature is constant and equal in both pipelines. To achieve Reynolds number similarity between the two pipelines, find the ratio of flow rate of water Q_w to flow rate of air Q_a.

Bibliography

Buckingham, E. "On Physically Similar Systems: Illustration of the Use of Dimensional Equations," *Phys. Rev.*, 4, 345–376, 1914.

Franck, C ,Waite, L. "Mathematical Model of a Variable Aperture Mitral Valve," *Biomed. Sci. Instrum.*, 38, 327, 2002.

Munson, BR, Young, DF, Okiishi, TH, *Fundamentals of Fluid Mechanics,* John Wiley & Sons, New York, 1994.

Szabo, G, Soans, D, Graf, A, Beller, C, Waite, L, Hagl, S. "A New Computer Model of Mitral Valve Hemodynamics during Ventricular Filling," *Eur. J. Cardiothor. Surg.*, 26, 239–247, 2004.

10

Lumped Parameter Mathematical Models

10.1 Introduction

Since the pioneering work of Otto Frank in 1899, there have been many types of mathematical models of blood flow. The aim of these models is a better understanding of the biofluid mechanics in cardiovascular systems. Mathematical computer models aim to facilitate the understanding of the cardiovascular system in an inexpensive and noninvasive way. One example of the motivation for such a model occurred when Raines et al. in 1974 observed that patients with severe vascular disease have pressure waveforms that are markedly different from those in healthy persons. By developing a model that would predict which changes in parameters affect those pressure waveforms in what way, the scientists might provide a means for diagnosing vascular disease before it becomes severe.

Lumped parameter models are in common use for studying the factors that affect pressure and flow waveforms. A lumped parameter model is one in which the continuous variation of the system's **state variables** in space is represented by a finite number of variables, defined at special points called **nodes**. The model would be less computationally expensive, with a correspondingly lower **spatial resolution**, while still providing useful information at important points within the model. Lumped parameter models are good for helping to study the relationship of cardiac output to peripheral loads, for example, but because of the finite number of lumped elements, they cannot model the higher spatial-resolution aspects of the system without adding many, many more elements.

Chapter 10 begins by addressing the general electrical analog model of blood flow, based on electrical **transmission line equations**. Later in Chap.10 a specific model of flow through a human mitral valve is presented in some detail.

10.2   Electrical Analog Model of Flow in a Tube

In Chap. 7, Sec. 7.7, a solution was developed, which was published by Greenfield and Fry in 1965, that shows the relationship between flow and pressure for axisymmetric, uniform, fully developed, horizontal, Newtonian, pulsatile flow. The Fry solution is particularly useful when modeling the relationship of the pressure gradient in a tube or blood vessel to the instantaneous blood velocity. In Chap. 8, we used the Fry solution to develop an electrical analog of a typical pressure-measuring catheter. Using well-known solutions to **RLC circuits**, it became possible to characterize parameters like the natural frequency and dimensionless damping ratio of the system.

I describe, here in Chap. 10, an alternative development of the modeling of flow through any vessel or tube based on that analog. Let the subscript V represent properties of the blood vessel that we are modeling and the subscript L represent properties of the terminating load. In Chap. 8, the terminating load was a transducer, but in the more general case of the isolated blood vessel within a cardiovascular system, the terminating load represents the effects of a group of distal blood vessels, usually arterioles and capillaries.

Figure 10.1 shows the electrical schematic of a circuit representing a length of artery, terminating in a capillary bed. The values for resistance, capacitance, and inductance for each resistor, capacitor, and inductor, respectively, are calculated from the blood vessel properties on a per unit length basis.

Recall that the hydrodynamic resistance of a blood vessel depends on its radius and length as well as the viscosity of the fluid flowing in the

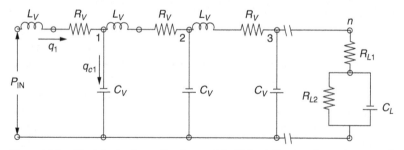

Figure 10.1   Electrical schematic of a model of blood flowing through a vessel or tube, with the V transcript representing the characteristics of the vessel and L representing the terminal load.

tube. For this type of **discretized model**, the resistance in each discrete resistor can be written as

$$R_V = \ell R_{\text{viscous}} = c_v \frac{8\mu\ell}{\pi r^4} \tag{10.1}$$

Using the SI system of units, the dimension of R_V is Ns/m^5. The inductance in the vessel for each discrete inductor element becomes

$$L_V = \ell L = c_u \frac{\rho\ell}{\pi r^2} \tag{10.2}$$

The inductance will have units of Ns^2/m^5. The capacitance of the vessel is the compliance of the vessel, or dA/dP, and depends on the pressure at the point where the capacitor is located.

$$C_V = \frac{q}{dP/dt} \tag{10.3}$$

Its units in SI are m^5/N. Note that the resistance, inductance, and capacitance of each vessel segment can vary, for example, in a tapering vessel.

10.2.1   Nodes and the equations at each node

In electrical circuit analysis, a node is defined as a point in a circuit where two or more **circuit elements** join. In Fig. 10.1, we will take the combination of a **resistor, inductor,** and **capacitor** as being a single element of our model, and number the nodes at the center of each element at the points where the resistor, capacitor, and inductor are joined. The nodes in Fig. 10.1 are labeled 1, 2, 3, . . . , n.

In the model, the input flow to the first element is labeled q_1. The input pressure is labeled P_{in}. The flow leaving node 1 toward the capacitor is labeled q_{c1}. The pressure at node 1 is P_1. Note that the model does not contain any information about the pressure at points between the input point and node 1.

Considering node 1, we can now write an equation for the pressure drop across the first resistor and inductor as shown in Eq. (10.4) as the first-order differential equation

$$P_{\text{in}} - P_1 = q_1 R_{V1} + L_{V1} \frac{dq_1}{dt} \tag{10.4}$$

In Eq. 10.4, P_{in} represents the input pressure to the vessel. P_1 is the pressure at node 1 and q is the flow through the vessel. The time rate of change of flow is designated dq/dt. Note that L_{V1} and R_{V1} can vary from node to node, and can even vary with pressure, which would cause our model to be nonlinear.

A second differential equation, which is also first order, can be written at node 1. This equation describes the flow into the capacitor at node 1. The flow q_1 is the vessel compliance at this point C_1 multiplied by dP/dt as shown in Eq. (10.5). Conceptually, at a specific point in time, it may be useful at this point to think about pressure at each of the nodes, P_1, P_2, ..., P_n, as being the independent variables and the flows, q_1, q_2, ..., q_n, as being the dependent variables, which depend on pressure. Mathematically, the P's and q's are dependent on time and location, and time is the only independent variable. We will use the model to try to understand the relationship between flow and pressure in the vessel.

$$q_{C1} = C_{V1}\frac{dP_1}{dt} \tag{10.5}$$

This system of two first-order differential equations can be written once for each node in the model. At the end, we will end up with a system of $2n$ first-order, ordinary differential equations, with two first-order equations written for each node: node 1 through node n. Although a large model can be computationally complex, the equations for a single node are relatively straight forward.

If we repeat the two equations for node 2 we end up with the following:

$$P_1 - P_2 = q_2 R_{V2} + L_{V2}\frac{dq_2}{dt} \tag{10.6}$$

$$q_{C2} = C_{V2}\frac{dP_2}{dt} \tag{10.7}$$

The general equations for a general node i, between node 2 and node n, will be

$$P_{(i-1)} - P_i = q_i R_{Vi} + L_{Vi}\frac{dq_i}{dt} \tag{10.8}$$

$$q_{Ci} = C_{Vi}\frac{dP_i}{dt} \tag{10.9}$$

10.2.2 Terminal load

If we are modeling the aorta, for example, it is possible to divide the vessel into segments and write a set of equations for each finite segment. If we continue to add the downstream details of every branch of the aorta, then the model will become larger and larger, more and more complicated, and more and more computationally expensive. Finally, it becomes impractical to individually model each one of the tens of billions of capillaries in the circulatory system. Instead, a more practical solution is that we will lump together all of the elements downstream of, or distal to, the main vessel that we are trying to understand with our model.

Because the capillaries are primarily resistance vessels we could begin estimating a load resistance that is based on the pressure at node n, and the flow moving through the entire capillary bed. The total, **terminal load resistance**, R_T is equal to the pressure at node n divided by the total flow q_n, as shown in Eq. (10.10).

$$R_T = \frac{P_n}{q_n} \qquad (10.10)$$

Although R_T would be a good first-order estimate of the terminal load in our model, we also know, from empirical evidence, that the capillary bed does not act as a pure resistance element. If we ended the model of the aorta, for example, in a single resistance, we would find that pressure waves are reflected proximally because of a mismatch in impedance. In many cases, we would predict standing pressure waves by examining the model, where we know from the empirical data that no standing pressure waves truly exist. That is to say, the pressure along the vessel will not show a steady pressure gradient as one might expect, but a time-varying pressure gradient that looks like periodic noise when one plots pressure gradient versus distance along the axis of the vessel. In fact, the vessels downstream of our model that make up the terminal load also exhibit a capacitive effect.

It has been suggested that a terminal load for our model could be estimated as a resistor in series with a second resistor parallel to a capacitor as shown in Fig. 10.2. For steady flow, or at very low frequencies, the total load impedance is equal to the total load resistance, which in this case is the sum of the two resistors. The total terminal resistance R_T is equal to R_{L1} plus R_{L2}, the sum of the values of the two terminal load resistors.

One method of estimating the capacitance of the load C_L is by **impedance matching**. For practical, biological systems, we would expect the output impedance of our model to match the input impedance of the terminal load, so that wave reflections are minimized. The ratio of R_{L1} and

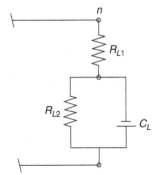

Figure 10.2   Terminal load of the model described in Sec.10.2.

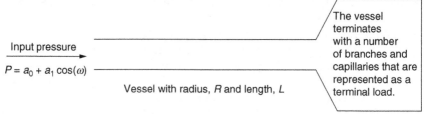

Input pressure

$P = a_0 + a_1 \cos(\omega)$

Vessel with radius, R and length, L

The vessel terminates with a number of branches and capillaries that are represented as a terminal load.

Figure 10.3   Conceptual model of simple electrical analog.

R_{L2}, as well as the value for capacitance, can be chosen to minimize wave reflections and to match the behavior of empirical data as closely as possible. The following example problem illustrates the electrical analog.

Example problem

To illustrate the use of the circuit analogy, we will model a single uniform section of an artery. The artery will be modeled as flexible, and it will be connected to a simulated end load. A steady state solution for pressure and flow rate will be calculated and interpreted.

The example problem will contain a general, i.e., symbolic approach. A special case using physical data will be included. The system is represented schematically in Fig. 10.3. The electrical analog is the circuit shown in Fig. 10.4.

The pressure imposed at the upstream end of the vessel being modeled is P_0. The pressure distal to the segment of vessel is P_1. These are analogous to voltages in the circuit. Another pressure which will be used in the solution is the pressure drop across the fluid resistance of the vessel, P_{RV}.

For the sake of completeness, the differential equations modeling the system will be derived in some detail. The dependent variables will be the

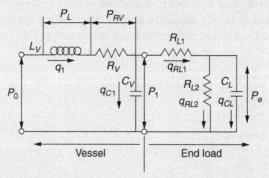

Figure 10.4   Electrical analog circuit for the example problem.

pressures just named. We will find a steady state solution for these pressures, and use it to further calculate the set of flow rates.

The solution will be expressed in the terminology of the electrical analog. The development uses the Kirchhoff voltage (KVL) and current laws (KCL). These are fundamental laws of circuit behavior. We supplement them with the equations describing the behavior of the circuit components, resistor, capacitor, and inductor. We begin with the first stage, giving the relationships pertaining to the components.

$$q_{C1} = C_V \frac{dP_1}{dt} \quad \text{the capacitor relationship} \tag{A}$$

$$P_L = L_V \frac{dq_1}{dt} \quad \text{where } P_L \text{ is the voltage drop across the inductor, } L_V \tag{B}$$

$$q_1 = \frac{P_{RV}}{R_V} \tag{C}$$

Next, we apply the KVL to the circuit, looping around the circuit elements and voltage drops in the stage.

$$P_0 - L_V \frac{dq_1}{dt} - P_{RV} - P_1 = 0$$

It is convenient at this point to substitute Eq. (C) into this equation.

$$P_0 - \frac{L_V}{R_V} \frac{dP_{RV}}{dt} - P_{RV} - P_1 = 0$$

Next, solve for the derivative.

$$\frac{dP_{RV}}{dt} = -\frac{R_V}{L_V} P_{RV} - \frac{R_V}{L_V} P_1 + \frac{R_V}{L_V} P_0$$

In this initial equation, P_0 is taken as an input, and we interpret P_{RV} and P_1 as two state variables. Another differential equation is needed to complete the description of the first stage.

We apply the KCL at the junction of the two stages.

$$q_1 - q_{C1} - q_{RL1} = 0$$

The currents are expressed in terms of voltages where P_{RL1} is the pressure drop across the load resistance R_{L1}

$$\frac{P_{RV}}{R_V} - C_V \frac{dP_1}{dt} - \frac{P_{RL1}}{R_{L1}} = 0$$

Again we solve for the derivative $dP_1/dt = 1/C_V[P_{RV}/R_V - P_{RL1}/R_{L1}]$. We now have a pair of linear first-order differential equations which describe the state of this part of the circuit. Please note that the second equation contains a voltage, P_{RL1}, from the next stage. This is to be expected since the stages are coupled and do not act independently.

Let us now look at this final piece of the circuit. We apply the KCL at the node inside the end load and find,

$$q_{RL1} - q_{RL2} - q_{CL} = 0$$

$$\frac{P_1 - P_e}{R_{L1}} - \frac{P_e}{R_{L2}} - C_L \frac{dP_e}{dt} = 0$$

Solve for the derivative, and thus obtain,

$$\frac{dP_e}{dt} = \frac{1}{C_L R_{L1}} P_1 - \frac{R_{L1} + R_{L2}}{C_L R_{L1} R_{L2}} P_e$$

Finally, we note that $P_{RL1} = P_1 - P_e$, and so we update the last equation in the second stage,

$$\frac{dP_1}{dt} = \frac{1}{C_V R_V} P_{RV} - \frac{1}{C_V R_{L1}} P_1 + \frac{1}{C_V R_{L1}} P_e$$

The differential equations are now presented in their finalized form.

$$\frac{dP_{RV}}{dt} = -\frac{R_V}{L_V} P_{RV} - \frac{R_V}{L_V} P_1 + \frac{R_V}{L_V} P_0$$

$$\frac{dP_1}{dt} = \frac{1}{C_V R_V} P_{RV} - \frac{1}{C_V R_{L1}} P_1 + \frac{1}{C_V R_{L1}} P_e$$

$$\frac{dP_e}{dt} = \frac{1}{C_L R_{L1}} P_1 - \frac{R_{L1} + R_{L1}}{C_V R_{L1} R_{L2}} P_e$$

For some economy of expression, we switch to the use of matrix notation at this point. Let

$$[A] = \begin{bmatrix} -\dfrac{R_V}{L_v} & -\dfrac{R_V}{L_v} & 0 \\[2ex] \dfrac{1}{C_V R_V} & -\dfrac{1}{C_V R_{L1}} & \dfrac{1}{C_V R_{L1}} \\[2ex] 0 & \dfrac{1}{C_L R_{L1}} & -\dfrac{R_{L1} + R_{L2}}{C_L R_{L1} R_{L2}} \end{bmatrix}$$

and

$$\{b\} = \begin{bmatrix} \dfrac{R_V}{L_V} \\ 0 \\ 0 \end{bmatrix}$$

It is convenient to accompany them with a set of relationships for the currents so that these can be output at any solution point. These relationships are as follows.

$$\begin{bmatrix} q_1 \\ q_{RL1} \\ q_{C1} \\ q_{RL2} \\ q_{CL} \end{bmatrix} = \begin{bmatrix} \dfrac{1}{R_V} & 0 & 0 \\ 0 & \dfrac{1}{R_{L1}} & \dfrac{1}{R_{L1}} \\ \dfrac{1}{R_V} & -\dfrac{1}{R_{L1}} & \dfrac{1}{R_{L1}} \\ 0 & 0 & \dfrac{1}{R_{L2}} \\ 0 & \dfrac{1}{R_{L1}} & -\dfrac{R_{L1} + R_{L2}}{R_{L1}R_{L2}} \end{bmatrix} \begin{bmatrix} P_{RV} \\ P_1 \\ P_e \end{bmatrix} \quad \text{or} \quad \{Q\} = [R]\{P\}$$

The steady state solution of these equations will now be described. We have calculated the matrix $[A]$ and the vector $\{b\}$ describing this linear system. Let the harmonic forcing function be represented by one harmonic at a time, such as $a_n \cos(n\omega t)$. (In this example, there is but one harmonic term, so $n = 1$.) The differential equation is

$$\frac{d}{dt}\{P\} = [A]\{P\} + \{b\}a_1 \cos(\omega t)$$

A solution of the following form is the particular or steady state solution to the differential system.

$$\{P\} = \{C_1\} \sin(\omega t) + \{C_2\} \cos(\omega t)$$

To find the vectors of unknown constants, $\{C_1\}$ and $\{C_2\}$, we first differentiate the proposed solution with respect to time, and then substitute this result into the differential equation.

$$\omega\{C_1\} \cos(\omega t) - \omega\{C_2\} \sin(\omega t)$$

$$= [A]\{C_1\} \sin(\omega t) + [A]\{C_2\} \cos(\omega t) + \{b\}a_1 \cos(\omega t)$$

Next, simply equate the sine and cosine terms on each side of the equal sign.

$$\omega\{C_1\} = [A]\{C_2\} + \{b\}a_1$$

and

$$-\omega\{C_2\} = [A]\{C_1\}$$

This leads to

$$\left(\omega I + \frac{1}{\omega}[A]^2\right)\{C_1\} = \{b\}a_1$$

This is a system of linear equations, which is solved for $\{C_1\}$, and then a simple matrix multiplication yields $\{C_2\}$.

$$\{C_1\} = \left(\omega I + \frac{1}{\omega}[A]^2\right)^{-1}\{b\}a_1$$

$$\{C_2\} = -\frac{1}{\omega}[A]\{C_1\}$$

This solution tells us the expected fact that the steady state solution associated with this harmonic will have the same frequency as that harmonic, but will experience a phase difference from it, since both sine and cosine terms are present.

Special case.   The data are presented in the following table.

Radius, R	8	mm
Length, L	10	cm
Viscosity, μ	0.004	Ns/m²
Density, ρ	1,050	kg/m³
Mean pressure, a_0	13,300	Pa
Fluctuating press, a_1	2,670	Pa
Frequency of fluctuation, first harmonic	6.28	rad/s

1. Calculation of the **Mean Flow Solution.**

For this vessel, we adjust the total fluid resistance in the terminal load ($R_T = R_{L1} + R_{L2}$) so, under pressure a_0 of 13.3 kPa, we get a steady flow of 4.0 L/min, corresponding to 6.65×10^{-5} m^3/s. The fluid resistance that gives this result is 2×10^8 Ns/m^5. The Reynolds number based on the mean flow across the cross section and also averaged over the time period of 1390 is high, since we have flow through a major vessel, but is not high enough so that the flow begins to leave the laminar range, even at peak velocity.

2. Calculation of the Parameters for the **Time-Varying Solution**.

We calculate the Fry parameters associated with the first **harmonic**, as described in Chap. 7. For the resistance per length and inertance per length of the first harmonic, we use the following formulas.

$$r_V = c_v \frac{8\mu}{\pi R^4} \qquad R_v = L r_V$$

$$l_V = c_u \frac{\rho}{\pi R^2} \qquad L_V = L l_V$$

The results are laid out in the following table.

Angular frequency	ω	6.28 rad/s
Womersley number	α	10.2
Resistance coefficient	c_v	2.22
Inertance coefficient	c_u	1.13
Resistance	R_V	5.529×10^5 Ns/m^5
Inductance	L_V	5.927×10^5 Ns^2/m^5

We select a capacitance, $C_V = 2 \times 10^{-10}$ m^5/N as a typical compliance for a large blood vessel.

3. Calculation of the Parameters for the Terminal Load.

Speaking in the context of the electrical circuit analog, for a blood vessel having resistance per length R_V, inertance per length of L_V, and compliance (capacitance) C_V, the characteristic impedance is

$$Z_V = \left(\frac{R_V + j\omega L_V}{j\omega C_V} \right)^{1/2}$$

This result is found in the theory of electrical transmission lines. The end load can be shown to have the following impedance.

$$Z_e = R_{L1} + \frac{R_{L2}(1 - j\omega R_{L2} C_L)}{1 + R_{L2}^2 C_L^2 \omega^2}$$

Reflections in a transmission line result from the difference between the impedance in the line at the junction and the connected terminal load. Since reflections, or returning waves, are not pronounced in the circulatory system, we model the interaction between the vessel and the terminal load by matching the two impedances, setting

$$Z_e = Z_V$$

Inserting the values previously tabulated, and recalling that $R_{L2} = 2 \times 10^8 - R_{L1}$, the real and imaginary parts of the impedance equality become

$$1.726 \times 10^8 = R_{L1} + \frac{2 \times 10^8 - R_{L1}}{1 + 39.44(2 \times 10^8 - R_{L1})^2 C_L^2},$$

(Units will be Ns/m^5)

and

$$-1.275 \times 10^7 = \frac{(2 \times 10^8 - R_{L1})(-1.25587 \times 10^9 + 6.28 R_{L1}) C_L}{1.3944(2 \times 10^8 - R_{L1})^2 C_L^2}$$

(Units will be Ns/m^5)

The physically meaningful solution to the pair of equations is: $R_{L1} = 1.667 \times 10^8$ Ns/m^5, $R_{L2} = 3.33 \times 10^7$ Ns/m^5, and $C_L = 1.026 \times 10^{-8}$ m^5/N.

4. Calculation of the **Steady State Solution.**

For the reader who is following through these calculations, we first give the arrays [A] and {b}.

$$[A] = \begin{array}{|c|c|c|} \hline -0.9329 & -0.9329 & 0 \\ \hline 9043.2 & -29.94 & 29.94 \\ \hline 0 & 0.5836 & -3.511 \\ \hline \end{array} \qquad \{b\} = \begin{array}{|c|} \hline 0.9329 \\ \hline 0 \\ \hline 0 \\ \hline \end{array}$$

We use these to calculate the arrays {C$_1$} and {C$_2$}, leading to the following steady-state solution for the pressures.

$$\{P\} = \begin{bmatrix} -2.29 \\ 59.9 \\ 192 \end{bmatrix} \sin(6.28t) + \begin{bmatrix} 8.56 \\ 2670 \\ 102 \end{bmatrix} \cos(6.28t)$$

From this expression, with the use of the relationship $\{Q\} = [R]\{P\}$ we can calculate the flow rates. These rates are presented in the plot in Fig. 10.5.

Commentary. We complete the example with some brief comments. First, we note the presence of a flow, q_{C1}, which, in essence, is storage of fluid in the artery segment as it experiences pulsatile flow. A similar flow, q_{CL}, describes the storage and release of fluid as a function of time in the rest of the system. The latter of the flows seems not far out of phase with the flow through the vessel segment, while the former experiences a noticeable phase difference.

This model is intended to illustrate some of the most general procedures that are used in creating the electrical circuit analog. More elaborate models can be generated, for example, by adding multiple segments and branches. More elements per unit length can also be added to increase the spatial resolution of the model. These more complex models can make important contributions to the understanding of the behavior of the cardiovascular system. For example, a clinician may wonder which model parameters have the greatest effect on flow through a given vessel and well-designed and properly validated models can save enormous time and expense when compared with experiments in animal models.

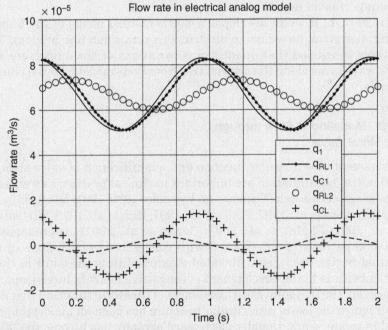

Figure 10.5 Flow rate results for example problem.

10.2.3 Summary of the lumped parameter electrical analog model

Many scientists have used similar models to mimic blood flow and pressure conditions in animal circulation as well as in the human circulatory system, with the goal of better understanding of the relationship between the two.

For example, Bauernschmidt et al. (1999a) simulated flow hemodynamics during pulsatile **extracorporeal perfusion**. Control of perfusion is achieved by surgeons, anesthesiologists, and perfusionists making real-time decisions. This leads to variations of the perfusion regimens between different hospitals and between different surgical teams in the same hospital. To develop a computer-controlled scheme for the integrated control of extracorporeal circulation, they developed a mathematical model for simulating hemodynamics during pulsatile perfusion. The model was constructed in Matlab using Simulink. They divided the human arterial tree into a 128-segment multibranch structure. Peripheral branches were terminated by resistance terms representing smaller arterioles and capillary beds. Bauernschmidt studied the effects of different perfusion regimens with differing amounts of flow and pulsatility and found the model to be useful for a realistic simulation of different perfusion regimens.

In 2004, L. R. John developed a mathematical model of the human arterial system, based on an electrical transmission line analogy. The authors concluded that quantitative variations of blood pressure and flow waveforms along the arterial tree from their model followed clinical trends.

10.3 Modeling of Flow through the Mitral Valve

Assessment of ventricular function and quantification of valve stenosis and mitral regurgitation are important in clinical practice as well as in physiological research (Thomas and Weyman, 1989, 1991; Takeuchi et al., 1991; Thomas et al., 1997; Scalia et al., 1997; Rich et al., 1999; Firstenberg et al., 2001; DeMey et al., 2001; Garcia et al., 2001). Approximately 400,000 patients are diagnosed with congestive heart failure in the United States each year. Elevated diastolic filling pressure in these patients leads to the development of congestive heart failure symptoms (Jae et al., 1997). Noninvasive assessment of **diastolic function** that does not require the use of intracardiac pressure has been an important goal, and in recent years, Doppler electrocardiography has become the "diagnostic modality of choice" (Garcia et al., 2001) to assess diastolic function.

The goal of the research that created this model was development and validation of a mathematical model of flow through the **mitral valve**

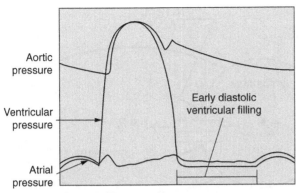

Figure 10.6 A typical pressure waveform in the left atrium, left ventricle, and aorta; the section that is marked shows the period designated as "early diastolic ventricular filling" or "E-wave filling."

during **early diastolic ventricular filling** (also sometimes referred to as **E-wave filling**), as shown in Fig. 10.6. This model may be used to assess diastolic left ventricular function based on **Doppler velocity waveforms** and cardiac geometry.

A secondary goal of the project was to publish representative values of important input parameters including effective mitral valve area, transvalvular inertial length, blood viscosity, blood density, atrial compliance, ventricular compliance, left ventricular active-relaxation characteristics, and initial pressure and flow values in the porcine model, which might be of use to other scientists.

By comparing the computer model output values to the measured pressure and flow rate data in pigs, parameters such as atrial compliance, ventricular compliance, and effective area were estimated. This is an inversion of the process used in some earlier papers in which pressure and flow predictions proceed from impedance and compliance data.

The model that was developed, which I describe in this section, calculates the flow and pressure relationships between the left atrium and the left ventricle through the mitral valve during early ventricular filling prior to atrial systole. I developed this model with the help of some students and in collaboration with cardiac surgeons at the University of Heidelberg Heart Surgery Laboratory. It was first published in the *European Journal of Cardiothoracic Surgery* in September 2004 (Szabo et al., 2004).

10.3.1 Model description

The model of early ventricular filling, or the E-wave portion of a single heartbeat, begins at the time when pressures are equal in the atrium and the ventricle; i.e., it begins from the instant of mitral valve opening.

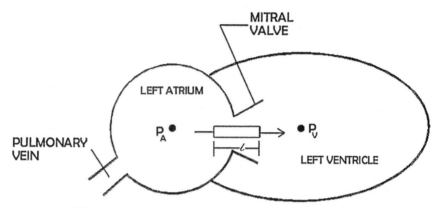

Figure 10.7 Schematic of the left heart used to develop a mathematical model of flow through the mitral valve. P_A is atrial pressure and P_V is ventricular pressure.

The model then describes flow and pressures during ventricular filling up to the point of atrial systole. Figure 10.7 shows a schematic of the atrium and ventricle with the mitral valve in-between. A fluid cylinder of length L is shown.

A system of three ordinary differential equations, Eqs. (10.11) to (10.13), describes the system. For the entire model, I will use the following representations:

P_a = left atrial pressure P_v = left ventricular pressure

q = instantaneous flow rate R_v = viscous resistance
  through the mitral valve
                 M = inertance term
R_c = convective resistance
                 C_v = ventricular compliance
C_a = atrial compliance
                 V = average velocity of blood through
A = mitral valve effective area   the mitral valve

R_v, R_c, M, C_a, C_v are described in greater detail in the text following.

The first differential equation describing flow rate through the mitral valve is shown in Eq. (10.11).

$$\frac{dq}{dt} = \frac{(P_a - P_v - R_v q - \text{sgn}(q)\, R_c q^2)}{M} + V\frac{dA}{dt} \qquad (10.11)$$

In the electrical analog, pressures are analogous to voltages in the circuit diagram. Flows are analogous to currents. The convective resistance term is analogous to a resistor, whose resistance is dependent on the current, which flows through it. Convective resistance will be discussed in more detail in Sec. 10.3.3. The term $V\frac{dA}{dt}$ represents the time rate of change of flow rate due to the closing or opening of the mitral valve. If the valve were held completely open, this term would be zero. Section 10.3.4 describes, in greater detail, the variable-area mitral valve that is used in this model.

The second differential equation is written as follows:

$$\frac{dP_a}{dt} = \frac{-q}{C_a} \tag{10.12}$$

where C_a represents atrial compliance in $\text{m}^5/\text{N}\left(\frac{\text{m}^3}{\text{N/m}^2}\right)$. The pressure inside the atrium is related to its volume through atrial compliance, $C_a = d(\text{volume})/dt$. The volume is also related to flow rate. In one sense, the flow into the atrium determines its final volume and therefore the final atrial pressure.

Similarly, the third equation describing the system is the equation for change in ventricular pressure with time. The ventricular pressure is similarly related to the ventricular volume and the ventricular compliance, C_v. The change in pressure of the ventricle is also related to "active relaxation." The active relaxation term, resulting from ventricular geometry, is described in more detail in Sec. 10.3.2.

$$\frac{dP_v}{dt} = \frac{q}{C_v} - \text{active relaxation term} \tag{10.13}$$

In all three equations, q represents flow rate in m^3/s, dq/dt represents the time rate of change of flow rate in m^3/s^2, P_a represents left atrial pressure in N/m^2, P_v represents left ventricular pressure in N/m^2, and M represents the inertance term, which is analogous to inductance in the electrical circuit and has the units of kg/m^4.

M in Eq. (10.14) can be calculated as

$$M = \rho l/A \tag{10.14}$$

where ρ represents blood density in kg/m^3, l represents the blood column length through the mitral valve with units of meters, and A represents the mitral valve effective area in m^2.

R_v in Eq. (10.15) represents viscous resistance, which is analogous to resistance in the electrical circuit and has the units $\text{kg/m}^4\text{s}$. R_v can be calculated as

$$R_v = 8 \rho l \mu/A^2 \tag{10.15}$$

where μ represents viscosity with units of Ns/m^2.

R_c represents convective resistance, which, in an electrical circuit model, would be analogous to a resistor whose variable resistance is dependent on current through the resistor, and has the units kg/m^7. R_c can be calculated as

$$R_c = \rho l / A_2 \tag{10.16}$$

10.3.2   Active ventricular relaxation

Another parameter affecting the final ventricular pressure during filling is active relaxation. To imagine active relaxation think of an empty plastic 1-L soft drink bottle. If you squeeze the bottle, the volume decreases, forcing air out of the bottle. Now when you release the bottle because of its geometry the bottle will expand, and if air is not allowed to flow back into the bottle (place your hand over the entrance) the pressure inside it will decrease below atmospheric. At the beginning of diastole, the left ventricle expands in a similar way and we say that active ventricular relaxation is present.

For the purpose of the model, active relaxation can be measured empirically and included in the model as a term, which decreases the pressure in the ventricle. When active relaxation occurs, it looks like a step change of pressure in a near critically damped system. Active relaxation parameters include a time constant for the pressure drop off and also a ratio of the maximum atrial pressure (pressure just prior to active relaxation) to minimum atrial pressure.

10.3.3   Meaning of convective resistance

To help understand the meaning of convective resistance, consider the Bernoulli equation while neglecting viscous resistance. The transmitral pressure drop, ΔP, is equal to $^1/_2\,\rho V^2$ where V is the transmitral velocity. R_c represents the convective resistance or resistance to flow associated with acceleration due to spatial changes rather than temporal changes. Blood must speed up as it passes through the mitral valve, because the valve acts like a nozzle on a garden hose. By using a nozzle on a garden hose, you can increase the velocity of the water as it exits the hose without increasing the pressure driving the water. In the same way that water is speeded up when leaving a garden hose using a nozzle, blood must be accelerated as it moves through the mitral valve. That acceleration is the explanation for convective resistance.

10.3.4   Variable area mitral valve model description

Previous lumped parameter models of diastolic filling modeled the mitral valve as a cylinder of a fixed cross-sectional area and length (Thomas and Weyman, 1989, 1992; Flachskampf et al., 1992; Waite et al., 2000).

Although the valve opening is clearly not constant since the valve by nature must open and close, it was possible to make a reasonable first estimate of the flow through the valve by modeling it as instantaneously 100 percent open or 100 percent closed.

In the current version of the model, the valve is modeled with a time-varying mitral opening area. One piece of the lumped parameter model described earlier is a model of the intrinsic and extrinsic characteristics of the valve itself, disregarding the heart chambers. We will call this portion of the model the "variable-area mitral valve model." A complete description of the variable area mitral valve model that was used as a part of this model can be seen in Franck and Waite (2002).

By using a systems approach, the mitral valve aperture can be viewed abstractly as a mechanical system whose behavior is governed by intrinsic dynamics and the forces acting on it. The intrinsic dynamics of the valve aperture are modeled by a second-order linear differential equation. This way of modeling takes into account the mass of the valve cusps, the elasticity of the tissue, and the damping experienced by the valve cusps while it also helps to keep the valve model simple. The equation already contains two of the six parameters the valve model uses: the damping coefficient D and the natural frequency of the mitral valve ω. D describes the amount of damping the valve cusps experience, which mainly depends on blood viscosity and the size of the cusps. The natural frequency ω takes into account the mass of the valve cusps and the elasticity of the tissue.

$$\frac{1}{\omega^2}\ddot{A} + \frac{2D}{\omega}\dot{A} + A = F(t) \tag{10.17}$$

The intrinsic dynamics of the forces acting on the valve, $F(t)$, are represented by the following equation with the parameters explained later in Sec. 10.3.5.

$$F(t) = (A_{\max} - A)\left[K_s(P_a - P_v) + K_d sgn(V)V^2 + K_a\frac{dV}{dt}\right] \tag{10.18}$$

Equation (10.17) is a second-order, nonlinear differential equation. The variable area A replaces all occurrences of the constant effective mitral valve area in the earlier model.

10.3.5 Variable area mitral valve model parameters

The equation for the variable area introduced six new parameters into the model. (1) $A_{\max}$ is the cross-sectional aperture of the mitral valve when fully opened. (2) The Greek character ω represents the natural frequency of the mitral valve in s^{-1}. It is influenced by the mass of the valve cusps and the modulus of elasticity of the tissue. (3) D represents the damping coefficient of the mitral valve. The valve cusps have comparatively little mass, but have a large area and they move in blood, a viscous fluid. Therefore, the system will possess significant damping ($D > 1$). The viscosity of the blood and the size of the valve cusps govern the magnitude of the damping coefficient. The damping coefficient has to be chosen so that no overshoot occurs; i.e., the actual mitral valve area must never be greater than the maximum area. (4–6) K_s, K_d, and K_a: These ("gain factors") represent the sensitivity of the valve to static, dynamic, and acceleration-induced pressure. K_s must be unitless, while K_d has the unit mass/volume and K_a has the unit mass/area.

10.3.6 Solving the system of differential equations

The model described here in Sec. 10.3 uses the Matlab command ode45 to solve the system of ordinary differential equations. The command ode45 is based on an explicit Runge-Kutta formula, the Dormand-Prince pair. It is a one-step solver. That means that when computing y at some time step n, it needs only the solution at the immediately preceding time point, y at time point $(n - 1)$. Matlab recommends ode45 as the best function to apply as a "first try" for solving most initial value problems.

10.3.7 Model trials

The model was validated by collecting empirical data on pigs at the University of Heidelberg, Heart Surgery Laboratory (Szabo et al., 2004). When using a mathematical model with many parameters, it becomes an important task to determine which parameters will be considered inputs and which will be outputs. If we have 100 parameters in the model, and 99 are considered inputs, and 1 is considered the output of the model, we need to decide a priori which parameter is most important to select.

Primary inputs to the model during the Heidelberg study were the diastolic atrial and ventricular pressure curves, a parameter describing left-ventricular active relaxation, mitral valve geometry, blood density, and blood viscosity. For these trials, atrial and ventricular compliance can also be considered model inputs. In other situations, it might be useful to estimate the values for these parameters from the model. The primary output of the model was calculated flow across the mitral valve. The waveforms of experimental and calculated flow across the mitral valve were compared to validate the model.

10.3.8 Results

Table 10.1 shows the various parameter values used for modeling the six porcine trials. The numbers in column 3 represent the data used in the trial shown in Fig. 10.8. In that figure, typical results measured in

TABLE 10.1   List of the Various Parameter Values Used for Modeling the Six Porcine Trials

	1	2	3	4	5	6	Average
α_a, cm^{-3}	0.26	0.079	0.09	0.05	0.06	0.029	0.095
α_v, cm^{-3}	0.17	0.22	0.21	0.12	0.12	0.16	0.17
l, cm	1.5	2	1.5	1.5	1.5	1.7	1.6
K_a, g/cm^2	0.002	0.002	0.002	0.0005	0.001	0.001	0.0014
K_d, g/cm^3	0.003	0.002	0.0015	0.0012	0.0003	0.0017	0.0016
K_s	0.007	0.002	0.0015	0.0001	0.0009	0.001	0.0021
Max valve aperture, cm^2	2	0.8	1.7	1.7	1.8	0.95	1.5
ω, rad/s	30	30	30	30	30	25	29.2
D, g/s	10	10	10	10	10	5	9.2
MSV, cc	16.6	9.6	11.4	12.8	11.3	4.9	11.1
SSV, cc	10.0	10.1	10.5	14.1	11.6	8.8	10.9

trial 3 in a porcine model are shown. The figure shows that when the atrial and ventricular pressures are correctly modeled and closely match the experimentally measured pressures, the calculated flow of blood through the mitral valve also matches the experimentally measured value. This also means that the measured stroke volume also compares closely to the stroke volume predicted by the model.

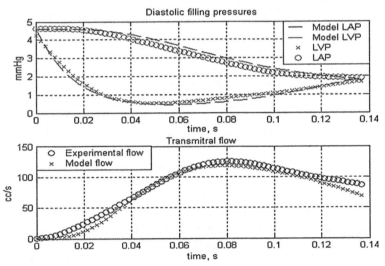

Figure 10.8 Pressures and flow curves from both model and porcine trials. LVP = left ventricular pressure; LAP = left atrial pressure. The different parameter values for this particular trial are listed in column 3 in Table 10.1.

The following parameters were held constant in all simulations; time duration of simulation = 0.137 s, integration step size = 0.0049 s, offset = 0 mmHg, μ = 0.035 dyne s/cm^2, ρ = 1 g/cc. MSV denotes measured stroke volume from data and SSV denotes simulation stroke volume, both have units cubic centimeter.

Table 10.1 shows the MSV compared to the SSV. The data demonstrate that flow through the mitral valve in the porcine model can be modeled successfully with comparable stroke volumes. By matching flow and pressure waveforms, it was possible to predict an unknown parameter like atrial compliance.

10.4   Summary

Chapter 10 presents an introduction to lumped parameter models and shows a few examples of their uses. Research that leads to a better understanding of the cardiovascular system in humans is important, but also problematic. In general, because of ethical considerations, it is not possible to perform research on humans simply to gain a better understanding of how the body functions. Even if we are able to simulate some aspects of human systems with animal research, that type of research is very expensive and can even be wasteful, and inhumane in the worst of cases. The development of mathematical models that can noninvasively simulate important parameters in human systems is a lofty goal and one that should be embraced.

On the other hand, without proper validation in appropriate animal or human systems, these models are sometimes little better than educated guesses. Chapter 10 introduces a mathematical model of flow that is based on equations similar to those of the transmission line equations in electrical engineering. Examples of published uses of similar models are shown.

Finally, in Chap. 10, I present a more in-depth overview of a lumped parameter model of flow through the mitral valve. The model is based on a system of nonlinear ordinary differential equations that are solved in Matlab using the ode45 command. The chapter presents some typical results from a porcine model and shows some promising results from that model.

Approximately 400,000 patients are diagnosed with congestive heart failure in the United States each year. Elevated diastolic filling pressure in these patients leads to the development of congestive heart failure symptoms. Noninvasive assessment of diastolic function that does not require the use of intracardiac pressure has been an important goal. In recent years, Doppler electrocardiography has become the "diagnostic modality of choice" to assess diastolic function. The goal of the model described in Chap. 10 was development and validation of a mathematical model of flow through the mitral valve during early diastolic ventricular filling (also sometimes referred to as E-wave filling), which promises assessment of diastolic left ventricular function using non-invasively collected Doppler waveforms.

Review Problems

1. Using Matlab or a computer algebra system, set up and repeat the calculations of the example problem following Sec. 10.2.2.

2. Try a new version of the example problem above with a different set of data. See the table.

Radius, R	8.5	mm
Length, L	15	cm
Viscosity, μ	0.0042	Ns/m^2
Density, ρ	1,050	kg/m^3
Mean pressure, a_0	12,500	Pa
Pulse press, a_1	2,500	Pa
Frequency of fluctuation, first harmonic	7.5	rad/s

3. Extend the example by adding two additional stages. Derive the matrices [A], {b}, and [R]. Solve the example problem, but with an artery that has taper. Let the segments have radii of 8.0, 7.5, and 7.0 mm. Assume that the length is still 10 cm. Keep the "capacitance" constant.

Bibliography

Bauernschmitt R, Naujokat E, Mehmanesh H, Schulz S, Vahl CF, Hagl S, Lange R. "Mathematical Modelling of Extracorporeal Circulation: Simulation of Different Perfusion Regimens," *Perfusion*, 14(5), 321–330, 1999a.

Bauernschmitt R, Schulz S, Schwarzhaupt A, Kiencke U, Vahl CF, Lange R, Hagl S. "Simulation of Arterial Hemodynamics after Partial Prosthetic Replacement of the Aorta," *Ann. Thorac. Surg.*, 68(4), 1441–1442, 1999b.

De Mey S, De Sutter J, Vierendeels J, Verdonck P. "Diastolic Filling and Pressure Imaging: Taking Advantage of the Information in a Colour M-mode Doppler Image." *Eur J Echocardiogr. Review.* 2(4):219–233, 2001.

Firstenberg MS, Greenberg NL, Smedira NG, McCarthy PM, Garcia MJ, Thomas JD. "Relationship between Systolic and Diastolic Function with Improvements in Forward Stroke Volume Following Reduction in Mitral Regurgitation." *Comput. Cardiol.* 28: 177–180, 2001.

Fisher J. "Comparative Study of the Hydrodynamic Function of Six Size 19 mm Bileaflet Heart Valves," *Eur. J. Cardiothorac. Surg.*, 9(12), 692–695, 1995; *discussion* 695–696.

Flachskampf F, Weyman A, Guerrero JL, Thomas JD. "Calculation of Atrioventricular Compliance from the Mitral Flow Profile: Analytic and in-Vitro Study," *J. Am. Coll. Cardiol.*, 19, 998–1004, 1992.

Franck C. *A Lumped Variable Model for Mitral Valve Aperture during Diastolic Filling* [M.S. thesis], Rose-Hulman Institute of Technology, Terre Haute, IN, 2001.

Franck C, Waite, L. "Mathematical Model of a Variable Aperture Mitral Valve," *Biomed. Sci. Instrum.*, 38, 327–331, 2002.

Garcia M, Firstenberg M, Greenberg M, Smedira M, Rodriguez L, Prior D, Thomas J. "Estimation of Left Ventricular Operating Stiffness from Doppler Early Filling Deceleration Time in Humans," *Am. J. Physiol. Heart Circ. Physiol.*, 280(2), H554–561, 2001.

Gorlin R, Gorlin SG. "Hydraulic Formula for Calculation of the Area of the Stenotic Mitral Valve, Other Cardiac Valves, and Central Circulatory Shunts, *Am. Heart J.*, 41(1), 1–29, 1951.

Greenfield JC, Fry DL. "Relationship between Instantaneous Aortic Flow and the Pressure Gradient," *Circ. Res.*, 17, 340–348, 1965.

Isaaz K. "A Theoretical Model for the Noninvasive Assessment of the Transmitral Pressure-Flow Relation," *J. Biomech.*, 25(6), 581–590, 1992.

Jae KO, Appleton CP, Hatle LK, Nishimura RA, Seward JB, Tajik AJ. "The Noninvasive Assessment of Left Ventricular Diastolic Function with Two-Dimensional and Doppler Echocardiography," *J. Am. Soc. Echocardiogr.*,10. 246–270, 1997.

John LR. "Forward Electrical Transmission Line Model of the Human Arterial System," *Med. Biol. Eng. Comput.*, 42(3), 312–321, 2004.

Lemmon J, Yoganathan AP. "Computational Modeling of Left Heart Diastolic Function: Examination of Ventricular Dysfunction, *J. Biomech. Eng.*, 122(4), 297–303, 2000.

Nudelman S, Manson A, Hall A, Kovacs S. "Comparison of Diastolic Filling Models and Their Fit to Transmitral Doppler Contours," *Ultrasound Med. Biol.*, 21(8), 989–999, 1995.

Raines JK, Jaffrins MY, Shapiro AH. "A Computer Simulation of Arterial Dynamics in the Human Leg," *Biomechanics*, 7, 77–91, 1974.

Reul H, Talukder N, Müller EW. "Fluid Dynamics Model of Mitral Valve," *J. Biomech.*, 14(5), 361–372, 1981.

Rich M, Sethi G, Copeland J. "Assessment of Left Ventricular Compliance during Heart Preservation," *Perfusion*, 13, 67–75, 1998.

Rich MW, Stitziel NO, Kovacs SJ. "Prognostic value of diastolic filling parameters derived using a novel image processing technique in patients > or = 70 years of age with congestive heart failure."*Am J Cardiol.* 1;84(1):82–86, 1999.

Scalia GM, Greenberg NL, McCarthy PM, Thomas JD, Vandervoort PM. "Noninvasive Assessment of the Ventricular Relaxation Time Constant (tau) in Humans by Doppler Echocardiography." *Circulation.* 95(1):151–155, 1997.

She JD, Yoganathan AP. "Three-Dimensional Computational Model of Left Heart Diastolic Function with Fluid-Structure Interaction," *J. Biomech. Eng.*,122(2), 109–117, 2000.

She J, Li M, Huan L, Yu Y. "Dynamic Characteristics of Prosthetic Heart Valves," *Med. Eng. Phys.*, 17(4), 273–281, 1995.

Szabo G, Soans D, Graf A, Beller C, Waite L, Hagl S. "A New Computer Model of Mitral Valve Hemodynamics during Ventricular Filling," *Eur. J. Cardiothorac. Surg.*, 26, 239–247, 2004.

Takeuchi M, Igarashi Y, Tomimoto S, Odake M, Hayashi T, Tsukamoto T, Hata K, Takaoka H, Fukuzaki, H. "Single-Beat Estimation of the slope of the End-Systolic Pressure-Volume Relation in the Human Left Ventricle," *Circulation*, 83, 202–212,1991.

Thomas J. "Prognostic Value of Diastolic Filling Parameters Derived Using a Novel Image Processing Technique in Patients > or = 70 Years of Age with Congestive Heart Failure," *Am. J. Cardiol.*, 84(1), 82–86, 1999.

Thomas JD, Weyman AE. "Numerical Modeling of Ventricular Filling," *Ann. Biomed. Eng.*, 20, 19–39, 1992.

Thomas JD, Weyman A. "Fluid Dynamics Model of Mitral Valve Flow: Description with in-Vitro Validation," *J. Am. Coll. Cardiol.*, 13, 221–233, 1989.

Thomas JD, Weyman A. "Echocardiographic Doppler Evaluation of Left Ventricular Diastolic Function," *Circulation*, 84(3), 977–990, 1991.

Thomas JD, Newell J, Choong C, Weyman A. "Physical and Physiological Determinants of Transmitral Velocity: Numerical Analysis," *Am. J. Physiol.*, 260, H1718–H1730, 1991.

Thomas JD, Zhou J, Greenberg N, Bibawy G, Greenberg N, McCarthy P, Vandervoort P. "Physical and Physiological Determinants of Pulmonary Venous Flow: Numerical Analysis," *Am. J. Physiol.*, 5(2), H2453–2465, 1997.

Verdonck P, Segers P, Missault L, Verhoeven R. "In-Vivo Validation of a Fluid Dynamics Model of Mitral Valve M-mode Echocardiogram," *Med. Biol. Eng. Comput.*, 34(3), 192–198, 1996.

Waite L, Schulz S, Szabo G, Vahl CF. "A Lumped Parameter Model of Left Ventricular Filling-Pressure Waveforms," *Biomed. Sci. Instrum.*, 36:75–80, 2000.

Index

Page numbers followed by *t*, *f*, or *n* indicate tables, figures, or notes, respectively.

CPSIA information can be obtained at www.ICGtesting.com
Printed in the USA
BVOW06*0934301015

424624BV00004B/121/P